American Institute of Chemical Engineers

Topical Conference on Transport Processes in Emerging Energy Systems 2004

AA003472

Held at the 2004 AIChE Spring National Meeting

April 25-29, 2004
New Orleans, Louisiana, USA

Printed from e-media with permission by:

Curran Associates, Inc.
57 Morehouse Lane
Red Hook, NY 12571
www.proceedings.com

ISBN: 978-1-60423-476-3

Some format issues inherent in the e-media version may also appear in this print version.

American Institute of Chemical Engineers

Topical Conference on Transport Processes in Emerging Energy Systems
2004

TABLE OF CONTENTS

Paper 32a

A Radioactive Tracer Study of Gas and Liquid Distribution in Horizontal CEN Paraxylene Unit Feed/Effluent Heat Exchangers

Jeffrey A. Amelse **

bp Amoco Chemical Company
150 West Warrenville Road
Naperville , IL 60566
AmelseJA@BP.com

** To whom correspondence should be sent

Steven D. Norwood
bp Amoco Chemical Company
P.O. Box 2215
Decatur, AL 35609

William Mixon
Tru-Tec Services, Inc.
8181 GSRI Road
Baton Rouge, LA 70820

Keywords: heat, exchanger, fouling, radioactive, tracer, paraxylene

Prepared for Presentation at the 2004 Spring National Meeting, New Orleans, LA, April 25-29
UNPUBLISHED

Abstract

The bp Amoco Chemical Company No. 2 Paraxylene Unit at Decatur, Alabama (DC2) was commissioned in 1978. It was designed for old naphthene-pool catalyst that operated at high H2/Hydrocarbon (H2/Hc) mole ratios. It has been debottlenecked extensively over the years, and now operates with a modern bp Amoco Chemical Company HSDE catalyst that operates at much lower H2/Hc mole ratio. The unit has its original eight feed-effluent exchangers. There are four trains of two exchangers in series. They are CEN double segmental no-tubes-in-windows designs. These are very large exchangers (6 foot diameter by 51 feet long). Preheated isomerization feed liquid is mixed with hydrogen rich recycle gas at the exchangers. Liquid is sprayed onto the tube sheet via six spray nozzles. It must make it into 1662 tubes in each exchanger. The in-series exchangers are stacked. Thus, unvaporized liquid leaving the bottom exchanger must be lifted via two phase flow and redistributed to the tubes of the top exchangers. Two of the exchangers have been reconfigured to preheat recycle gas (the A train). A small fraction of liquid is fed to these exchangers to enhance heat transfer, and take the load off of the other exchangers (the B/C/D trains), which are in true feed/effluent service.

These exchangers experience tube-side fouling, and fouling is most severe at the bottom of the exchangers. We suspected poor liquid distribution leading to localized very low H2/Hc mole ratios. A radioactive tracer study confirmed this hypothesis.

Ar-41 was injected into the recycle gas, and Br-82 labeled dibromobenzene was injected into the liquid. The study showed that:

1. The gas is uniformly distributed, but the liquid is poorly distributed.
2. The H2/Hc mole ratio is 3 times higher at the top of A1 than at the bottom.
3. The H2/Hc mole ratio is 1.5 times higher at the top of B1 than at the bottom.
4. Distribution of liquid to the 6 nozzles is uniform.
5. The dry point of the B/C/D trains is near the entrance to the hot shell.
6. The dry point of the A exchanger is beyond the entrance to the hot shell.

Background

Para-xylene (pX) is a bulk industrial chemical intermediate. It is oxidized and purified to Purified Terephthalic Acid (PTA). PTA is reacted with ethylene glycol to produce polyester, which has a number of uses, including fiber for clothing, film, tire cord, and plastic beverage bottles. World capacity for pX was about 24 million metric tons per annum in 2003. bp Amoco Chemical Company (bp) is one of the largest producers of pX.

The primary feedstock for pX is reformate mixed xylenes. Reformate mixed xylenes are separated from refinery reformate by distillation. The reformate mixed xylenes fraction is comprised primarily of the C8 aromatics (ethylbenzene (EB), and the three xylene isomers (ortho-, meta-, and para-)). The C8 non-aromatics have very close boiling points, and cannot be separated by distillation. Certain primarily C9 non-aromatics (paraffins and naphthenes (P&N's)) also boil in the range of the C8 aromatics. Thermodynamics favor higher conversion of non-aromatics in the naphtha reformer feed to aromatics at lower reformer pressure. Modern low pressure reformers can produce a mixed xylenes distillation cut with as little as about 0.5 wt% non-aromatics. Older medium or high pressure reformers produce a mixed xylenes distillation cut with as much as 6-10 wt% non-aromatics.

The xylenes are near their equilibrium distribution in reformate (about 25% pX, 50% mX, and 25% oX). The pX concentration in the mixed xylenes is further diluted by the presence of about 12-20% EB, and 0.5–10 wt% P&N's. Reformate mixed xylenes from older medium or high pressure reformers are often extracted (in the sulfolane extraction process licensed by UOP, for example). However, extraction is capital and energy intensive.

pX units based on reformate xylenes generally comprise three sections: isomerization, fractionation, and pX separation. pX is separated from mixed xylenes by one of two methods: low temperature crystallization or via selective adsorption on a molecular sieve. bp uses its own crystallization process.

The xylene isomers in the fresh mixed xylenes feed are generally near equilibrium. Thus, they are sent to the pX separation section directly or via the fractionation section. After pX removal, the pX separation section reject is sent to the isomerization section, where the xylenes are isomerized catalytically to near equilibrium. The catalyst must also convert EB and P&N's to lighter or heavier byproducts to prevent their build-up in the recycle loop, since they cannot be separated from pX by distillation.

All bp pX units use modifications of the Amoco (now bp) pX crystallization process, and bp xylene isomerization catalysts. The bp No. 2 pX unit at Decatur (DC2), Alabama, which was commissioned in 1978, is one of the largest pX units in the world. It was designed to process unextracted reformate mixed xylenes from the Amoco (now bp) refinery at Whiting, Indiana, which has medium pressure cyclic reformers. The Whiting mixed xylenes are sent to Decatur via pipeline. The DC2 unit was designed for an old naphthene pool-type catalyst that operated in the vapor phase, and required high hydrogen-to-hydrocarbon (H2/Hc) mole ratio on the order of 8:1. The unit never ran with that type of catalyst. Naphthene pool catalysts have largely been replaced by transalkylation- or hydrodeethylation-type catalysts following the discovery of medium pore molecular sieves in the 1970's. (For a discussion of xylene isomerization catalyst types and the impact of medium pore molecular sieves on their evolution, see Amelse, 1988a and 1988b). The DC2 unit currently uses a bp High Selectivity DeEthylation (HSDE) catalyst that converts EB primarily to benzene and ethane via reaction with hydrogen, and operates at very low H2/Hc mole ratios. The drop in mole ratio relative to the original design, together with various unit expansions over the years, means that the isomerization section feed/effluent heat exchangers are no longer designed for their current process conditions. The low mole ratio, high P&N's, and departure from design conditions lead to heat exchanger fouling.

The feed effluent exchangers are of CEN double segmental no-tubes-in-window (NTIW) design, to prevent tube vibration at the high design H2/Hc mole ratio. Isomerization section feed liquid is sprayed into the tubes, and carried via recycle gas. Poor liquid distribution is suspected to aggravate the fouling. This paper will discuss a radioactive tracer study of liquid and gas distribution in the exchangers.

Exchanger Configuration

The DC2 unit was designed with four trains of two horizontal CEN feed/effluent exchangers in series (the KE-104 A-D trains). These are big exchangers. They are 60 inch diameter by 51 feet in length. The hot shell is stacked above the cold shell as shown in Figure 1.

The bp Amoco Chemical Company processes incorporates proprietary heat integration between the isomerization and fractionation sections to reduce fuel consumption. To improve energy efficiency at the DC2 pX unit, the KE-104A train was re-piped to preheat the recycle gas stream. The A train receives all of the recycle gas, and a fraction of the isom feed liquid. The liquid is completely vaporized, and the recycle gas containing the vaporized liquid is then split to the B-D trains. There are six spray nozzles on each feed head. The double-segmental NTIW design leads to two strips of tubes as shown in Figure 2.

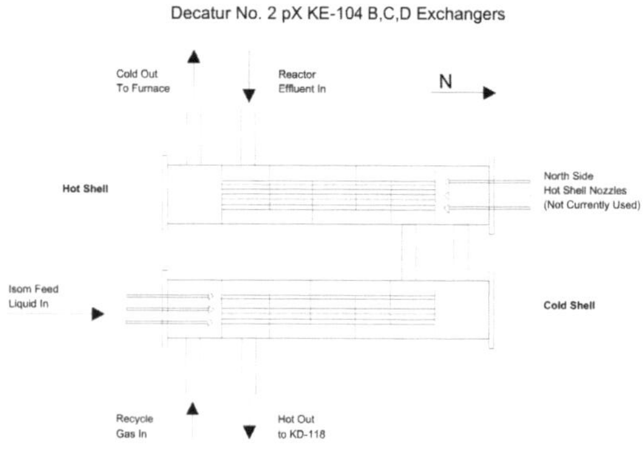

Figure 1. Exchanger Configuration

Figure 2. Exchanger Tubesheet

Experimental

40 cc of Argon-21 (half-life 1.8 hours) was used to probe gas distribution. About 2 cc of m-dibromobenzene was used to probe liquid distribution. Injection size was about 20-50 mCi for both isomers. These small quantities were all that was needed for injection into a liquid line flowing at greater than 400,000 kg/hr! Detectors were arranged as shown in Figures 3 and 4.

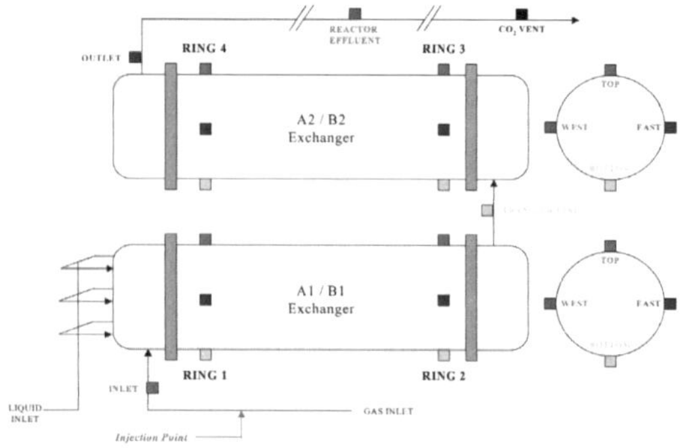

Figure 3. Gas Injection Detector Placement

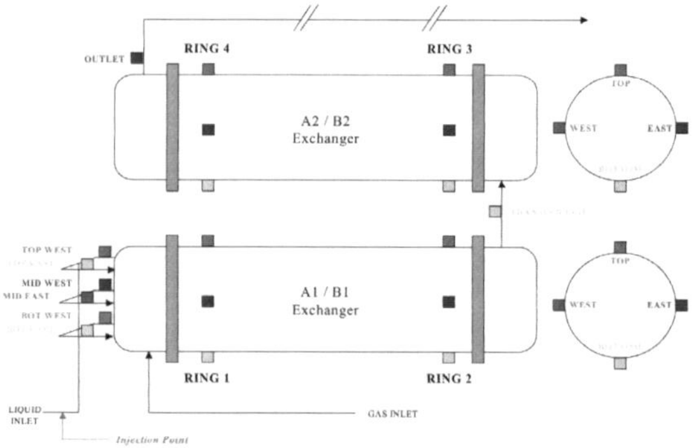

Figure 4. Liquid Injection Detector Placement

Results and Discussion

The Ar-41 gas injection indicates that gas distribution is essentially uniform in the exchangers. Signal strengths following gas injection is shown for the A-train Ring 1 (cold shell cold inlet) and Ring 4 (hot shell cold outlet) are shown in Figures 5 and 6, respectively. The signal has been attenuated by the time it hits Ring 4, indicating dispersion due to unequal tube residence time. In addition to information about gas and liquid distribution, the tracer study provides other useful information. Time between Ring 1 and Ring 4 peaks gives residence time in the exchanger. Time between successive peaks at the same detector gives residence time in the recycle gas loop (from compressor through the exchangers, furnace, reactor, etc., and back to the compressor). Signal strength at the top and bottom is higher than that at east and west, which is expected, since the top and bottom detector sees more tubes due to the tube strip configuration. However, top signal is essentially equal to bottom signal and east signal is essentially equal to west signal, indicating uniform gas phase distribution.

The liquid tracer injections indicate poor liquid distribution. Signal strength following liquid injection is shown for the A-train Rings 1-4 in Figures 7-10 respectively. While the A-train east and west signal strengths are comparable, the top signal is noticeably weaker than the bottom

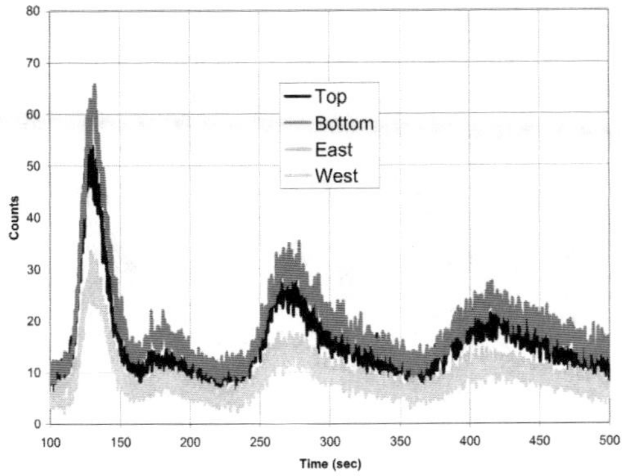

Figure 5. KE-104 A1 Gas Injection Ring 1

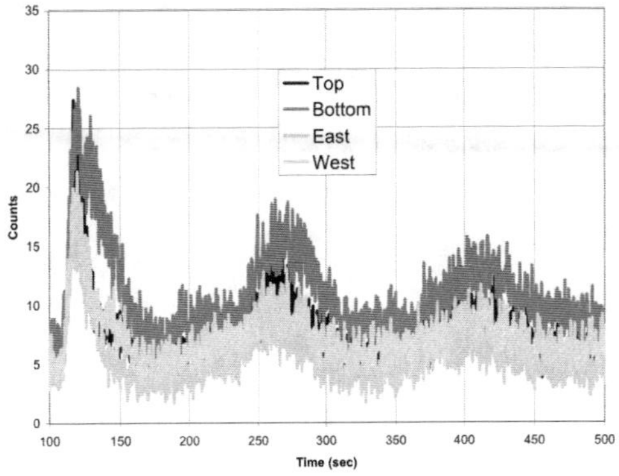

Figure 6. KE-104 A2 Gas Injection Ring 4

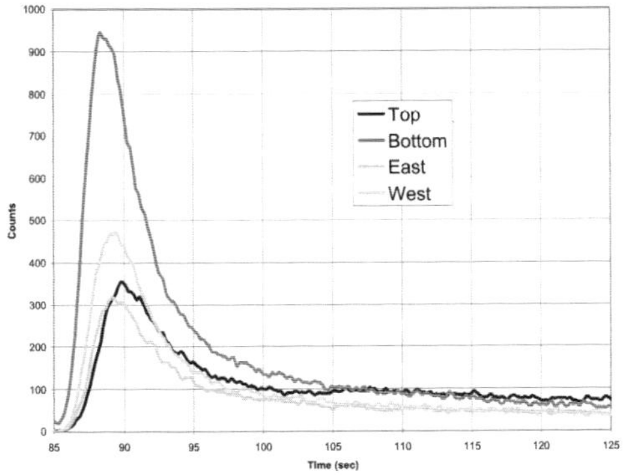

Fig. 7. KE-104 A1 Liquid Injection Ring 1

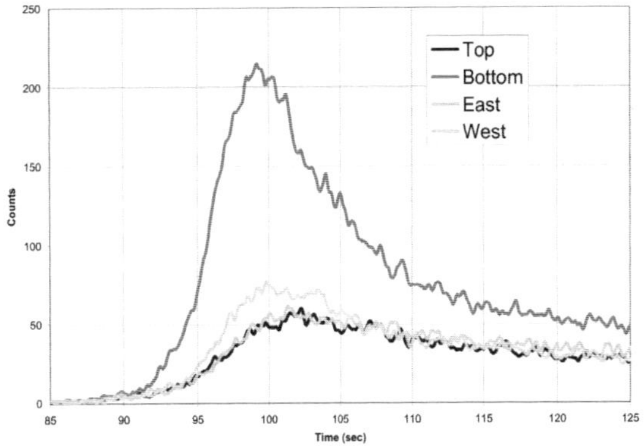

Fig. 8. KE-104 A1 Liquid Injection Ring 2

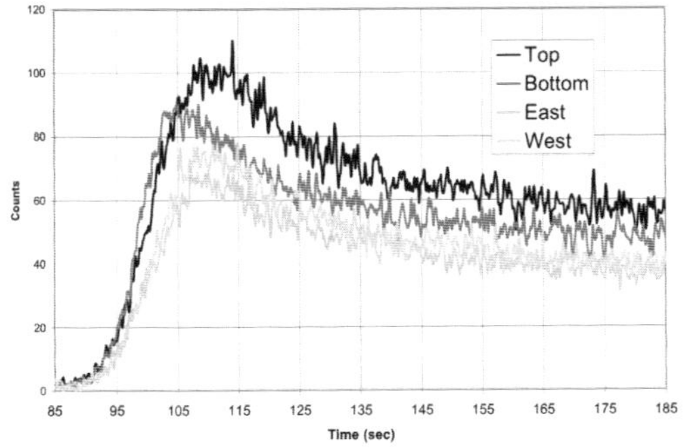

Fig. 9 KE-104 A2 Liquid Injection Ring 3

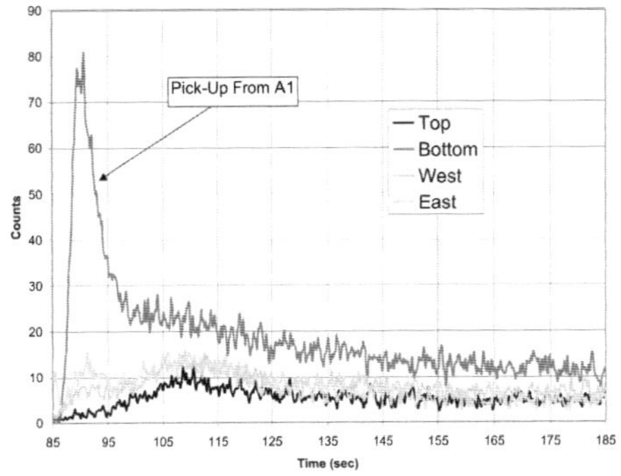

Figure 10. KE-104 A2 Liquid Injection Ring 4

signal. This indicates that more liquid is going through the bottom tubes. This leads to lower H2/Hc mole ratios at the bottom of the A1 exchanger and higher fouling. The liquid has largely redistributed by the time it hits the upper (hot) shell.

The normal boiling point of 1,3-dibromobenzene (218 C) is significantly higher than that of the C8 aromatics (mX n.b.p. = 139 °C). Thus, it lays down at the exchanger dry point. Very little signal is detected at Ring 4. Ring 3 signal strength shows a peak followed by a slow decay. This indicates the dry point is very near the entrance to the A2 hot shell.

Table 1. Relative Signal Strength Peak Areas

Exchanger	Ring	Top	Bottom	East	West
A1	1	28%	72%	43%	57%
	2	25%	75%	44%	56%
A2	3	52%	48%	48%	52%
	4	Barely Detectable – Tracer "stuck" in tubes			
B1	1	41%	59%	51%	49%
	2	40%	60%	52%	48%
B2	3	48%	52%	47%	53%
	4	Barely Detectable – Tracer "stuck" in tubes			

Liquid distribution is not as bad in the B-train. However, it is still far from uniform. Top:bottom signal strength for the B1 cold shell is 60:40. Once again, the liquid has largely redistributed by the time it has been lifted into the B2 hot shell.

It is easy to see why liquid distribution is poor after examining the photo of the tubesheet (Figure 2). Liquid is sprayed at the tubes through six nozzles. A good part of the liquid hits the blank area between the tube strips, and falls towards the bottom of the exchanger. This explains the higher bottom signal strength following liquid injection. The plant installed slanted angle irons on the tubesheet to try to redirect liquid that hits the blank area back into the tubes. However, this study shows this to be ineffective. We have recommended vertical baffles that extend from the tube sheet back beyond the nozzle heads to improve the situation. However, horizontal CEN exchangers with liquid spray nozzles are just not a good design for vaporizing service. Vertical vaporizers (Texas towers) became a better choice, followed by Packinox welded plate exchangers in which the liquid is sprayed up onto and lifted into the vertical places by high recycle gas velocity. The liquid inherently redistributes every few inches up the plates as it moves through chevron-shaped corrugations of the plate bundle, thus solving the liquid distribution problem.

REFERENCES

Amelse, J. A., Stud. Surf. Sci. Catal., Vol. 38, Proc. 10th N. Amer. Meeting of Catal. Soc., San Diego, CA May 17-22, 1987, J. W. Ward, ed., Elsevier, NY, p. 165 (1988a).

Amelse, J.A., 1988 AIChE Summer Natl. Meeting, Denver, CO, Aug. 21-24 (1988b).

PREHEAT TRAIN FOULING
PART I - EFFECT OF PROCESS PARAMETERS

C.B. Panchal, Patrick Malone, Scott Stomeirowski, and Ton W. Cao

Abstracts

The objective of this investigation was to determine the effects of process parameters on the rate of fouling in preheat train. A field-fouling unit was built and installed in preheat train at two refineries. Series of experiments with extended test period (>1,000 hours) were performed to determine the fouling rate at different operating conditions. The results were analyzed to determine threshold fouling conditions and identify individual heat exchangers of the preheat train operating above or below threshold fouling conditions. The field data were further validated with laboratory fouling results with the same crude.

Background

Crude oil distillation is the largest consumer of thermal energy among all distillation processes simply because it is the front-end separation of all petroleum processes, producing petroleum fractions that are fed to the downstream processing units. A simplified preheat train is shown in Figure 1. Crude is heated from an ambient temperature to about 260°C in preheat train, where a significant fraction, about 60%, of heat from product streams is recuperated. A fired heater provides the remaining heat to raise crude temperature to 340°C – 390°C prior to feeding the crude to the distillation column. The consequences of fouling of preheat exchangers are reduced heat recovery from products and a drop in the crude oil temperature at the fired heater inlet, resulting in increased fuel firing. If the heater reaches its safe firing limit then throughput must be reduced. A detailed analysis of an operating refinery by Panchal and HuangFu (2000) shows that fouling of preheat train exchangers increases energy use by more than 12,500 Btu/barrel of distillation feed, which is equivalent to 70 Trillion Btu/yr for the US refining industry. The analysis also showed that nearly one-half of this energy could be recovered by operating heat exchangers below threshold fouling conditions, physical and chemical mitigation, and/or optimized cleaning schedule.

The physical and chemical mechanisms associated with crude-oil fouling are complex, which hinder effective applications of mitigation methods. The absence of reliable predictive methods for the fouling rate and threshold fouling conditions, and the enormous costs of conducting field tests have led researchers to rely heavily on either laboratory fouling tests or experience from preheat train operation in formulating mitigation methods. The laboratory results and field experiences have shown that under certain conditions, the rate of fouling becomes relatively low. Panchal et al. (1999) investigated this concept of threshold fouling conditions and developed a correlation to predict threshold boundary-layer film temperature as a function of wall shear stress. This concept needs to be validated with a wide range of field experiments. This investigation focuses on obtaining controlled field data to determine the effects of process parameters on the rate of fouling and threshold-fouling conditions in preheat train. A field-fouling unit was built and installed in preheat train at two refineries. Series of experiments with extended test period (>1,000 hours) were performed to determine the fouling rate at different operating conditions. The results were analyzed to determine threshold fouling conditions, and identify individual heat exchangers of the preheat train operating above or below threshold-fouling conditions. The field data were further validated with laboratory fouling results with the same crude.

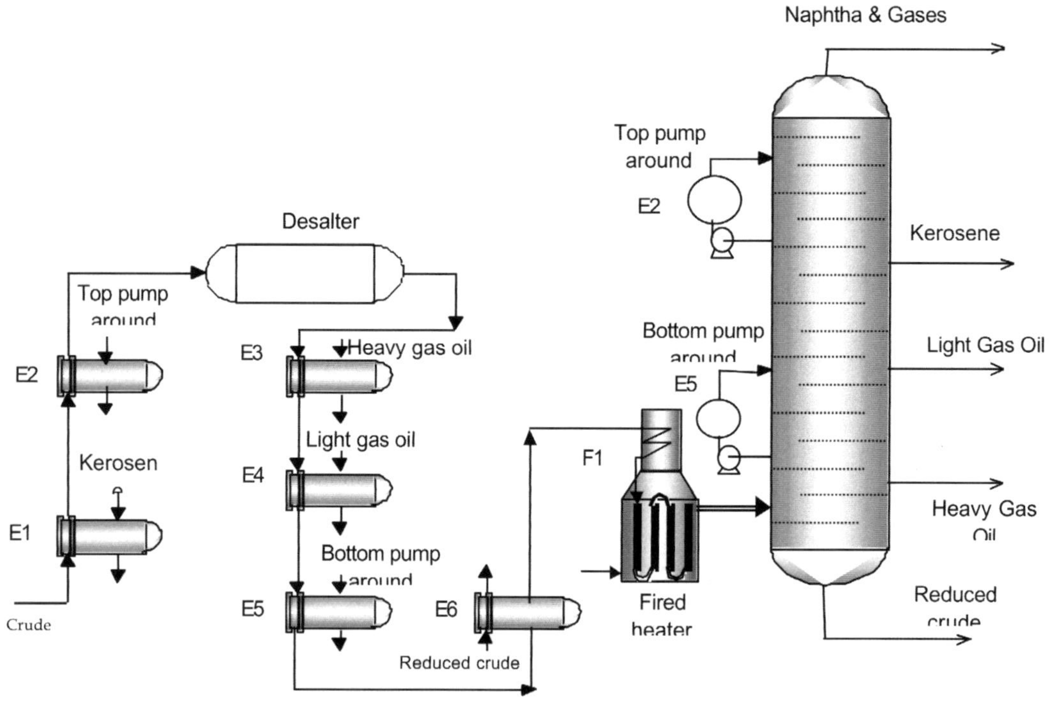

Figure 1. Simplified crude distillation preheat train

Fouling Experiments

Field experiments were conducted at two refinery sites. A field-fouling unit was designed to measure the fouling rate under once-though flow conditions using side stream of preheat train.
Figure 2 shows a photographic view of the field-fouling unit. The fouling unit was self-contained, skid-mounted, test rig with all the controls and instruments necessary to monitor fouling. The field unit consisted of two test sections, which provided the advantages of simultaneously monitoring the fouling rate at two sets of operating parameters and tube geometry. Each test section contained a Heat Transfer Monitor (HTM), which consisted of an assembly of two heating blocks installed on the tube by a shrink-fit method. Two sets of three thermocouples were installed in each block; one set of thermocouples at inlet and outlet ends of heated section. Each set of thermocouples provided temperature gradient data for the heating block. By measuring temperature gradients at inlet and outlet, an average value for the heat transfer coefficient for the test section was determined. The fouling resistance at time "t" was calculated by taking difference between the reciprocal of the heat transfer coefficient at time "t" and that at beginning of the test at which the probe surface is free of the fouling deposits.

Results and Discussion

Crude at field site A was medium, API ~33 with asphaltene content of 6 to 10%. Table 1 presents some of the test conditions and fouling characteristics. For tests 4 and 5, HTM1 was equipped with the enhance tube; the results are presented in Part II (Malone et al., 2004) of this paper. The fouling trend in general shows an induction period for a cleaned test section, after which the normal fouling rate is established. The effects of crude velocity and the surface temperature followed the trend observed in laboratory experiments at Argonne and at Westhollow Research Center. The results in Figures 3 and 4 show that the fouling rate is significantly reduced when the velocity was increased from 1 m/s to 1.4 m/s. The field data were also analyzed to determine threshold-fouling conditions.

Figure 2. Photographic view of the field-fouling unit

Table 1. Test conditions and fouling characteristics for site A

		Test period hours	Bulk temperature °C	Surface temperature °C	Velocity m/s	Fouling*
Test 1	HTM1	462	217	267	1.16	SF
	HTM2	462		257	0.98	SF
Test 2	HTM1	1159	218	242	1.16	SF
	HTM2	1159		247	0.98	SF
Test 3	HTM1	1600	229	324	1.37	NF
	HTM2	385		343	1.83	NF
Test 4	HTM2	450	214	270	0.61	F
Test 5	HTM2	1005	232	283	0.61	F

* F – high fouling rate; SF – slow fouling rate; NF – negligible fouling rate

Figure 5 presents threshold limits for the crude at Site A. Four sets of data points are presented in this threshold chart; 1) laboratory data obtained at Argonne; 2) laboratory data obtained at Shell Westhollow center, 3) data obtained by the field-fouling unit, and 4) three preheat train heat exchangers which were monitored in this investigation. The chart provides valuable design and operating guidelines to minimize fouling. The results show that two heat exchangers monitored in this investigation were operating above threshold conditions, and they showed high fouling rates. The third heat exchanger was operating close to the threshold-fouling condition, and it showed low fouling rate. However, at low throughput (i.e. lower velocity), the fouling rate would go up. The threshold chart also indicates that laboratory experiments, under proper simulated conditions, can provide valuable data to determine threshold-fouling conditions.

Crude at field site B was light, API > 40 with relatively low asphaltene content. This site showed fouling trend similar to that at site A, although the fouling mechanism was different. The performance of individual heat exchanger of the preheat train was analyzed and the fouling rate was determined for the duration of the test period. The resulting fouling rates are presented on the threshold-fouling chart, along with the fouling rate obtained with the field-fouling unit. The results are presented in the threshold-fouling chart in Figure 6. The results show that farther the heat exchanger operates above the threshold limit, higher the fouling rate is. Near and below the threshold limit, the fouling rate is relatively low or negligibly small.

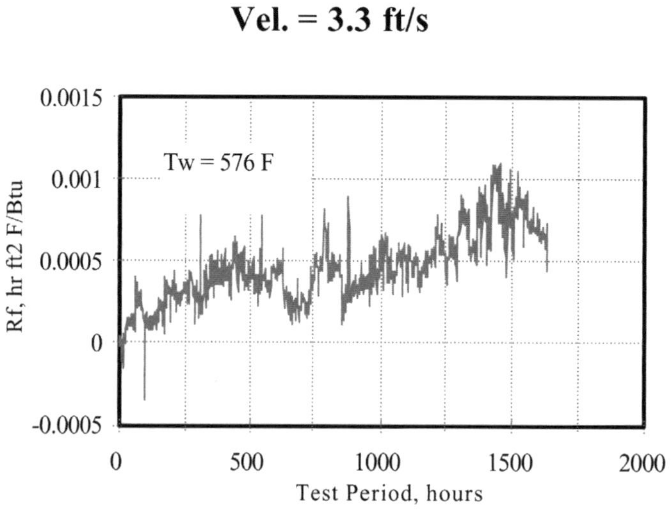

Figure 3. The fouling rate at Site A: velocity = 1 m/s; Ts=302C

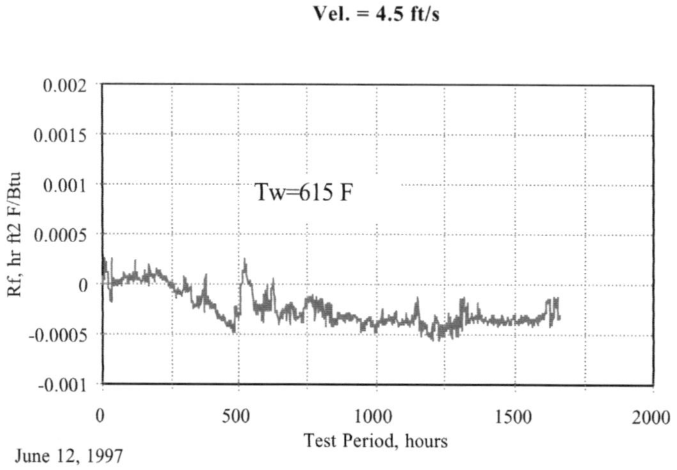

Figure 4. The fouling rate at Site A: velocity = 1.4 m/s; Ts=324C

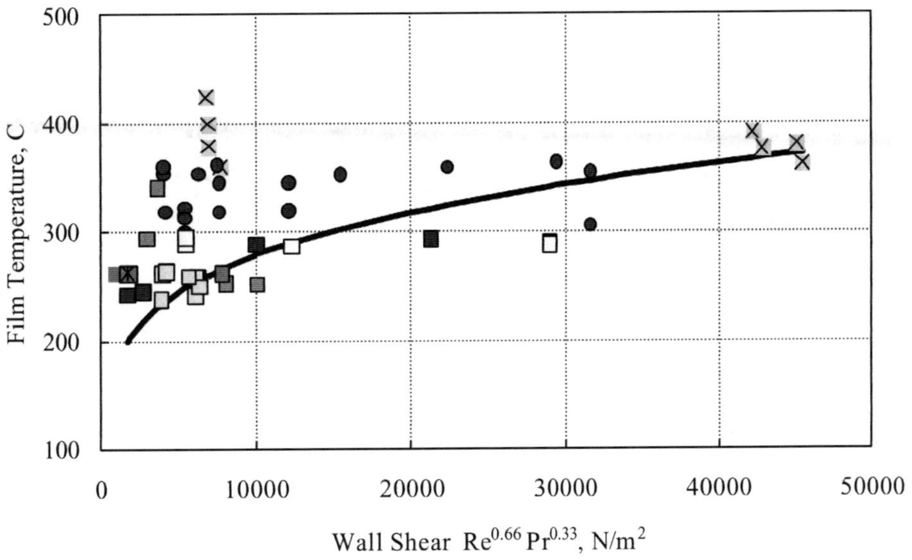

Figure 5. Threshold fouling chart for site A

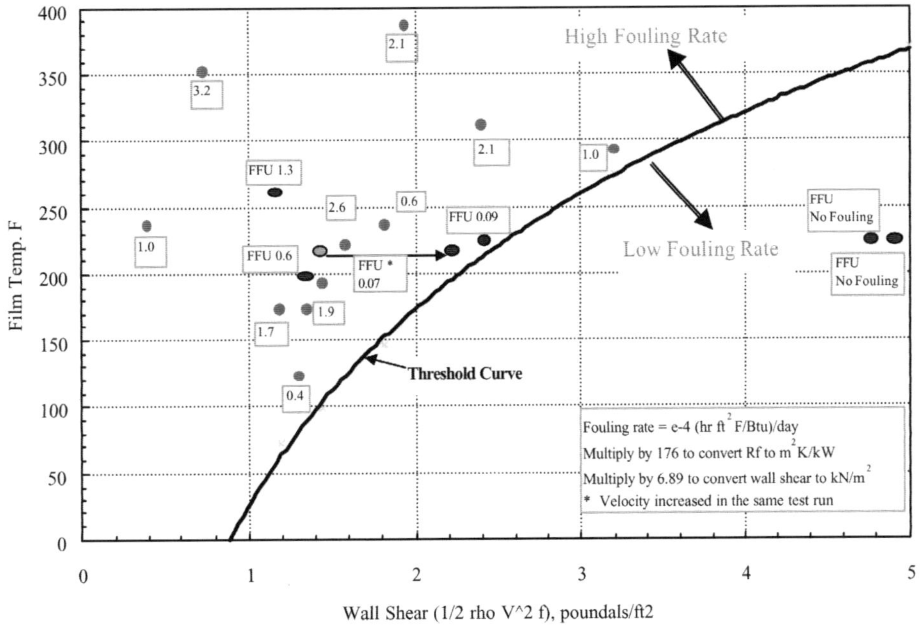

Figure 6. Threshold fouling chart for site B

Conclusions

The effects of crude velocity, crude blend, bulk and surface temperatures were determined. The fouling results obtained with the field-fouling unit were presented on the threshold-fouling chart; along with the fouling data obtained for the preheat train. The chart provides design and operating guidelines to minimize fouling by operating heat exchangers near or below threshold fouling limits.

Acknowledgement

This work supported by the U.S. Department of Energy, Assistant Secretary of Energy Efficiency and Renewable Energy, and the Office of Industrial Technology Program, under Contract W-31-109-Eng-38.

References

Panchal CB, Kuru WC, Liao CF, Ebert WA, and Palen JW, 1997, "Threshold conditions for crude-oil fouling," Proceedings of the Engineering Foundation Conference: *Understanding Heat Exchanger Fouling and its Mitigation*, Begell House Publisher, New York, NY.

Panchal CB and HuangFu EP, 2000, "Effects of mitigating fouling on the energy efficiency of crude-oil distillation," *Heat Transfer Engineering*, 21:3, pp. 3-9.

Malone P, Stomierowski S, Cao TW, Panchal CB, and Krueger A, 2004, "Preheat train fouling: Part II-mitigation by tube insert and enhanced tubes," Paper presented at the 2004 AIChE Spring Meeting, New Orleans, LA.

PREHEAT TRAIN FOULING
PART II - MITIGATION OF CRUDE-OIL FOULING BY TUBE INSERT AND ENHANCED TUBE

Scott Stomeirowski, Patrick Malone, Ton Cao, C.B. Panchal, and Arthur W. Kruger

Abstract

The objective of this investigation was to determine the effectiveness of wire insert and enhanced tube to reduce the fouling rate. A field fouling unit was built and installed on side stream of preheat train at two refineries for an extended test period. The fouling unit consisted of two test sections operating in parallel, which allowed direct comparison of the fouling rate of the enhanced tube and tube insert with the plain tube test section. Spirally indented tube was used as an enhanced tube and oscillating wire, an On-line Mechanical Cleaning Device (OMC), was used as tube insert. The results showed that under comparable operating conditions the tube insert and enhanced tube reduced the fouling rate more than one-half the plain-tube fouling rate.

Background

The industry has been coping with the fouling problem by cleaning heat exchangers on a routine basis and/or by applying chemical additives or installing physical devices on the basis of vendors' experiences. Part I paper (Malone et al., 2004) discussed the interactive effects of wall shear and the boundary-layer film temperature on the fouling rate, which is expressed by the threshold fouling correlation (Panchal et al., 1997). If operating conditions for a particular heat exchanger is not far from threshold conditions, then mitigation device can shift these conditions near threshold values. This will then reduce the fouling rate significantly and extend the run length of exchangers. If the fouling rate can be significantly reduced, cleaning can be scheduled at normal turnarounds.

The interactive effects of transport phenomena and fouling mechanisms with enhancement devices have been subject of investigation for different types of fouling (Watkinson, 1991; Rabas et al., 1991). The enhancement mechanisms alter the physical process of foulant formation and attachment to the heat-transfer surface. The resulting effects would be reduced fouling rate. When tube bundle cannot be changed, tube inserts can be installed in plain-tube exchangers to reduce the fouling rate. Two types of tube inserts are commercially available; 1) oscillating wire (Krueger, 2004), and 2) wire matrix (Gough, 2004). Whether tube insert or enhanced tube is used, it is important to understand the fouling mechanism to evaluate their effectiveness to reduce the fouling rate significantly to justify the added cost.

The objective of this investigation was to determine the effectiveness of oscillating wire insert and spirally indented enhanced tube to reduce the crude oil fouling rate. Fouling experiments were performed for an extended period of time to determine the effectiveness of these two physical mitigation devices under realistic refinery operating conditions.

Fouling Experiments

The field-fouling unit is described in Part I paper (Malone et al., 2004). A field fouling unit was built and installed on side stream of preheat train at two refineries. The fouling unit consisted of two test sections operating in parallel, which allowed direct comparison of the fouling rate of the enhanced tube and tube insert with plain tube. Spirally indented tube was used as an enhanced tube, and oscillating wire as On-line Mechanical Cleaning (OMC) device was used as tube insert. Figure 1 shows a photographic view of the test section equipped with spirally indented tube. This type of enhanced tube is commonly used in the refrigeration industry and in power plant condensers. The tube material was 410 stainless steel rather than copper-based alloys used in cooling-water applications. The indented surface disrupts the boundary layer and thereby enhances the heat transfer coefficient. In addition, it reduces the effective thickness of the thermal-boundary layer where fouling precursor(s) is generally formed. The combined effects of these two processes reduce the fouling rate, and in some cases make it negligibly small. Spirally indented tubes can be cleaned with conventional hydroblasting methods commonly used for plain tubes. In this investigation, the test section was brush cleaned with a brush similar to that used for cleaning plain-tube test section, and the performance was restored after each test run.

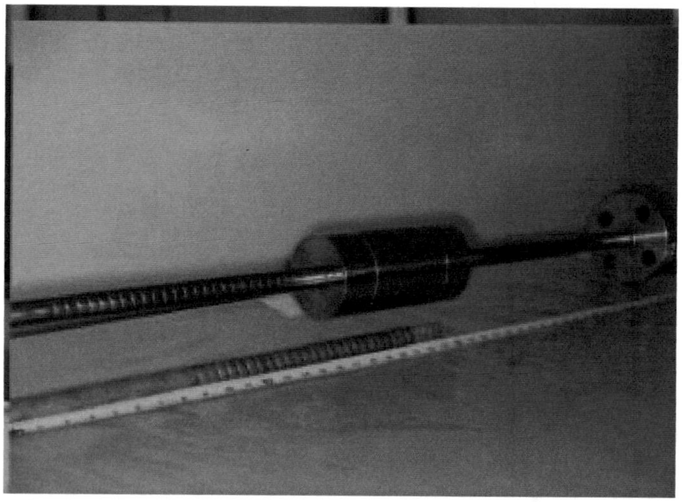

Figure 1. Test section equipped with spirally indented tube

Oscillating wire inserts have been installed in several industrial heat exchangers (Krueger, 2000). A schematic view of the wire insert is shown in Figure 2. This wire insert enhances the heat transfer coefficient, similar to the enhanced tube. In addition, the oscillating movement produces secondary flows in the vicinity of the tube surface, which enhances removal of fouling precursors formed in the boundary layer. Again, the combined effects of enhancing the heat transfer coefficient, which reduces the wall temperature, and the secondary flows reduce the fouling rate. The previous installations have been for heat exchangers operating at high Reynolds numbers with crude velocities greater than 1 m/s. Crude velocity in this investigation was less than 1 m/s.

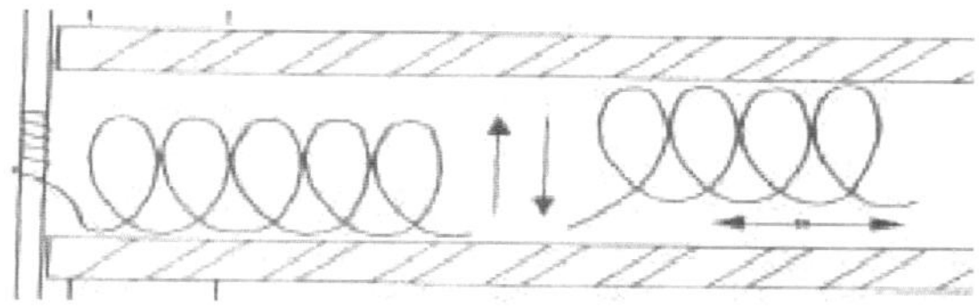

Figure 2. Schematic of oscillating wire insert

Crude at field site A was medium, API ~33, with asphaltene content of 6 to 10%, while crude at field site B was light, API > 40 with low asphaltene content. These two sites provided widely different crudes, specially asphaltene content, and hence fouling mechanisms.

Results and Discussion

Table 1 presents test conditions and fouling rates for spirally indented tube at Site A. The heat flux was maintained the same in Test 1, within 10%, for both test sections. Because the heat transfer coefficient for the enhanced tube is nearly twice the plain tube coefficient, the surface temperature is lower. The fouling rate for the plain tube was 0.022 (Km^2/kW)/day, whereas the fouling rate for the enhanced tube was negligibly small. In Test 2, the heat flux of the enhanced tube was increased, so that surface temperatures of both test sections were comparable. In this case, the fouling rate for the enhanced tube was about ½the fouling rate for the plain tube. In Test 1, the combined effects of lower surface temperature and increased wall shear stress caused the fouling rate to be negligibly small. In Test 2, the fouling rate was reduced due to increased wall stress and corresponding reduced thermal boundary layer. These two test runs clearly demonstrated the effectiveness of spirally indented tube to reduce the fouling rate.

Table 1. Test conditions and fouling characteristics of spirally indented tube at site A

	Test Period, hours	Bulk Temperature, °C	Surface Temperature, °C	Velocity, m/s	Fouling Rate, (Km2/kW)/day
Test 1: Same heat flux					
Plain tube	450	214	270	0.61	0.022
Enhanced tube			242		Negligible
Test 2 Same surface temperature					
Plain tube	1005	232	283	0.61	0.040
Enhanced tube			284		0.022

After completing initial tests with the enhanced tube, oscillating wire insert was installed in the plain-tube test section. The fouling experiment was then continued to run for an extended period (> 6,000 hours) of time. The results are shown in Figures 3 and 4, for the enhanced tube and oscillating wire, respectively. The fouling rates for both test sections were relatively low for the extended period of time.

The results for Site B are shown in Figures 5 and 6 for the plain tube and oscillating wire insert, respectively. The fouling trend of the plain tube reflects variations in process conditions including crude blending. The plain tube showed steady increase in the fouling resistance with each variation in the fouling trend. On the other hand, the fouling rate for the tube insert remained relatively low, after an initial increase in the fouling resistance.

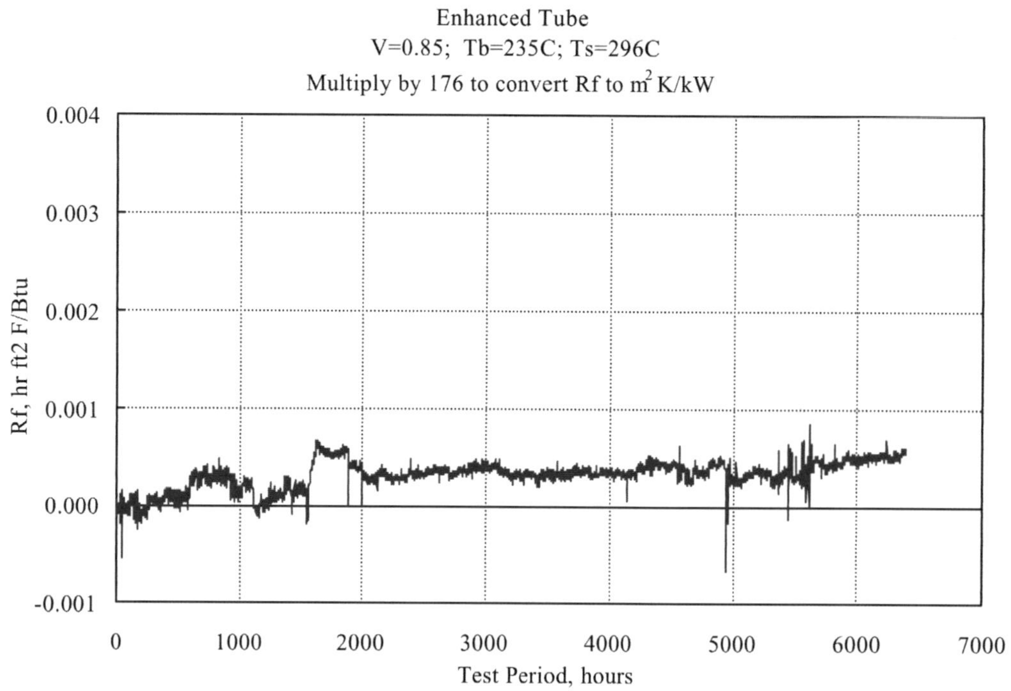

Figure 3. Fouling rate for enhanced tube Site A

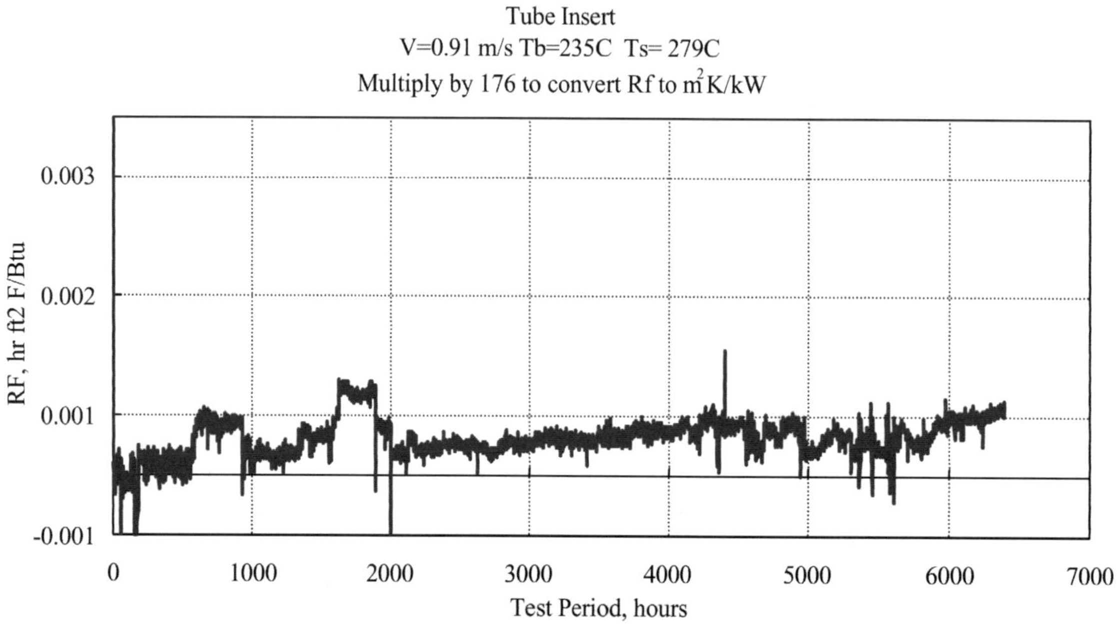

Figure 4. Fouling rate for plain tube with tube insert Site A

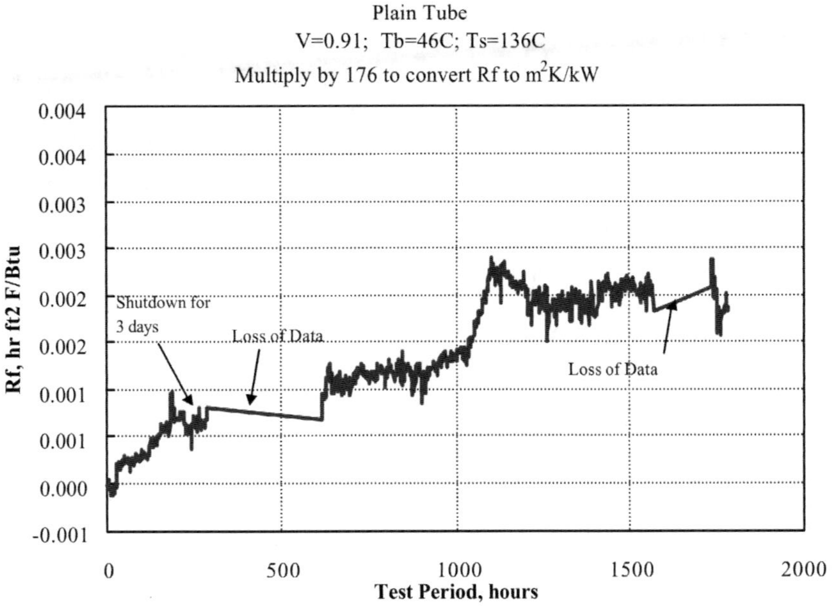

Figure 5. Fouling rate for plain tube Site B

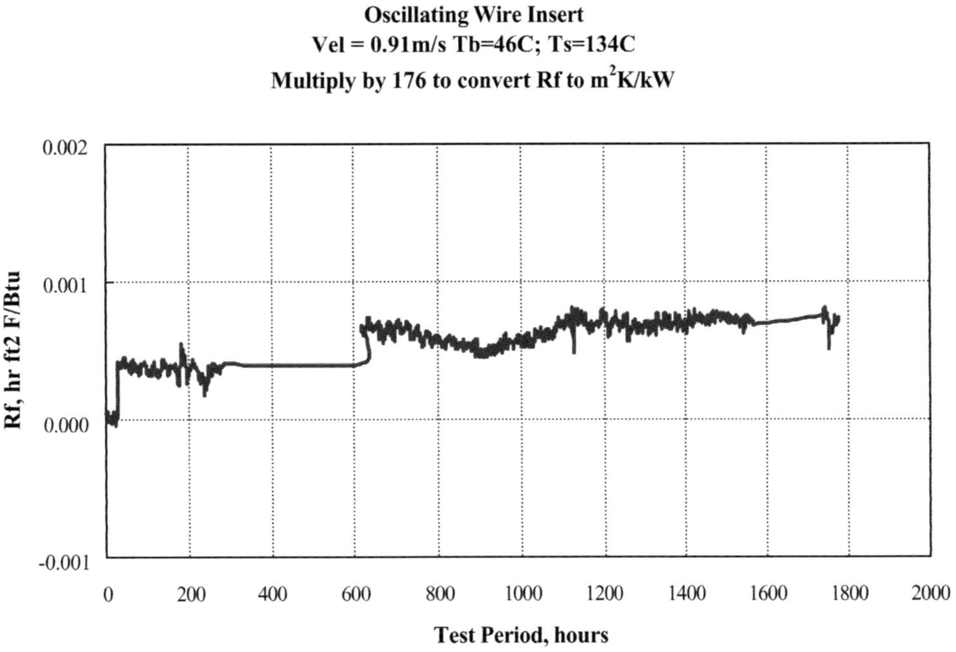

Figure 6. Fouling rate for plain tube with oscillating wire insert Site B

Conclusions

The effectiveness of spirally indented tube and oscillating wire insert was determined at two test sites, with widely different crudes and fouling mechanisms. The results showed that the fouling rate can be significantly reduced with the enhanced tube and oscillating wire insert. Application of these results to other industrial heat exchangers requires proper analysis of process conditions to determine threshold-fouling conditions and governing fouling mechanisms.

Acknowledgement

This work supported by the U.S. department of Energy, Assistant Secretary of Energy Efficiency and Renewable Energy, and the Office of Industrial Technology Program, under Contract W-31-109-Eng-38.

References

Droegemueller P, Smeethe A, Ellerby P, and Gough M, 2004, "The theory, historical application and current practice of using wire matrix elements to enhance heat transfer and reduce tube side fouling in hydrocarbon streams," Paper presented at the 2004 AIChE Spring Meeting, New Orleans, LA.

Krueger, A., 2004, "Enhancement of reboiler operational reliability in a chemical plant in Thailand by utilizing the "Spiref System" a Mechanical Online Cleaning (MOC) device," Paper presented at the 2004 AIChE Spring Meeting, New Orleans, LA.

Malone S, Stomieroski S, Cao TW, Panchal CB, 2004, "Preheat train fouling: Part I – effects of process parameters," Paper Presented at the 2004 Spring AIChE Meeting, New Orleans, LA.

Rabas TJ, Panchal CB, Sasscer DS, and Schaefer R, 1991, "Comparison of power-plant condenser cooling-water fouling rates for spirally indented and plain tubes," ASME Publication, HTD-Vol 164, pp. 29-38.

Watkinson AP, 1991, Interactions of enhancement and fouling," ASME Publication, HTD-Vol 164, pp. 1-8.

The theory, historical application and current practice of using wire matrix elements to enhance heat transfer and reduce tube side fouling in hydrocarbon streams

Martin J Gough[1], *Peter Droegemueller*[1], *Peter Ellerby*[1] *and Alex Smeethe*[1,2]

[1]*Cal Gavin Limited,*
Minerva Mill Technology Centre
Station Road, Alcester,
Warwickshire B49 5ET, England
Email: martin.gough@calgavin.co.uk
[2]*Department of Chemical Engineering, University of Birmingham,*
Edgbaston, Birmingham B15 2TT, UK

Wire matrix elements (hiTRAN® thermal systems) have progressively been used as an effective and economic way to improve the thermal performance of tubular heat exchangers and reactors for over 20 years. A conservative estimate of the number of exchangers fitted with this system, and used in a wide range of hydrocarbon services, would exceed 5000. Of this number many hundreds are in operation where the duty can be described as having a medium to high propensity to foul. Test data, plant operation and visual substantiation all point to boundary layer effects, reduced residence time of fluid at the wall together with related wall temperature changes, as being the primary benefit providers.

Introduction

Designing heat exchangers to satisfy the demands of thermal duty, economy of construction, pumping limitations, and longevity of operational life, remains a challenge for all engineers where the service is prone to fouling. The affect on profits associated with loss of performance, down time and maintenance are usually very significant to processing companies. In operation the need to accurately predict the fall off rate of heat exchangers in say a crude oil preheat train has seeded the development of highly sophisticated simulation programmes[1].

Extensive experimental work has also been carried out to evaluate the fouling mechanisms and potential fouling rates of a range of different hydrocarbons at varying bulk and wall temperatures and at different flow rates[2]. Varying pressure may also play a significant role, as does the mixing of different crudes and the presence of solids. Research typically involves the use of a heated probe inserted into a fluid stream, providing data on deposition rate. Whilst valuable knowledge is gained this way it is difficult to apply directly to the actual design of heat exchangers. The degree of complexity created by the many inter-dependant relationships set up by changes in physical, chemical, and operational circumstances makes design reliability very subjective.

In such circumstances limited tools are available to the designer to use process and equipment variables to good effect. Exchanger pass arrangements and the heating fluid temperature can usually be optimised although often limited in benefit. Alternatively tube inserts provide the option of modifying the fluid dynamics in-tube to affect the fouling mechanisms[3]. These changes include increased turbulence, increased local shear rates and reduced fluid residence time in contact with the tube wall. Also the enhancement of heat transfer coefficients provided by such devices can offer a real benefit through changing the wall temperatures, which is important in controlling fouling rates.

Dye stream visualisation

As a first step to understanding the flow patterns within hiTRAN® Elements, dye stream injection was used to visualise the effects. These flow patterns were found to be too complex to be solved analytically or using CFD.

Dye was injected at two points on the tube wall, at the top and bottom, as seen in figure 1. This was performed at a range of Reynolds numbers, with the results being captured with a video camera for later analysis. Prior to entry to a wire matrix element, the dye stream remained on the tube wall under laminar flow conditions, but immediately on encountering the first loop of the element, fluid was removed from the wall and mixed into the bulk flow.

Figure 1 Dye stream at Reynolds number 500 A: - plain tube conditions
B: - hiTRAN® matrix element (22mm diameter tube)

Laser Doppler measurement of flow patterns

Dual beam Laser Doppler Velocimetry (LDV) was used in order to quantify the effects of wire matrix elements on the fluid flow. The experiments were carried out over a Reynolds range of 100 to 10000 in a 20mm inner diameter glass tube. Data was collected for the plain tube case and for two different void fraction elements. Only a single velocity component was measured (down the tube) due to diffraction from the curved tube wall. All measurements were taken at a point 3mm from the end of the element.

$$\frac{U_x}{U_{mean}} = 2\left(1 - \frac{s^2}{r^2}\right) \quad \text{Equation 1}$$

$$\frac{U_x}{U_{mean}} = 1.22\left(1 - \frac{s}{r}\right)^{1/7} \quad \text{Equation 2}$$

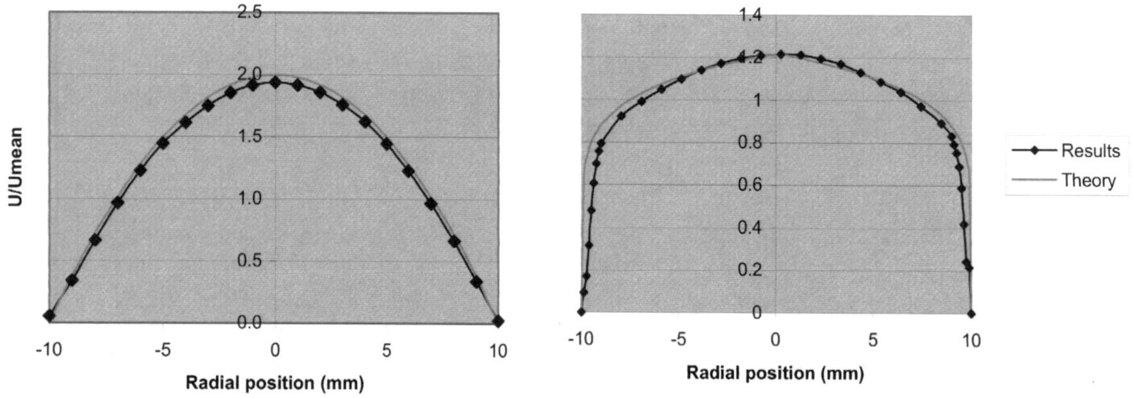

Figure 2 Velocity profiles measured by LDV compared with theoretical profiles for laminar (Left, Re100) and turbulent (Right, Re10000) for the plain tube case

A comparison of the measured results for the plain tube to the predictions from theory[4] (Equations 1& 2) can be seen in figure 2.
Good agreement was found between the measured and predicted results for the laminar (Reynolds number 100 and 500) cases, with an average deviation of less than 5%. A much larger deviation was found in the turbulent regime, though this was mostly near the wall. This deviation can be explained by the presence of a laminar boundary layer, slowing flow near the wall.

LDV measurements show that the fluid in a tube equipped with wire matrix elements flows in an annular fashion through the insert, with a 'dead zone' present in the centre due to the central core wire of the element (figure 3) It can be seen that significantly increased velocities near the tube wall were obtained by using these elements, resulting in a profile resembling the turbulent case near the wall (figures 3-4)

Further work is planned using PIV (Particle Imaging Velocimetry) in order to study the flow patterns around wire matrix elements in more depth.

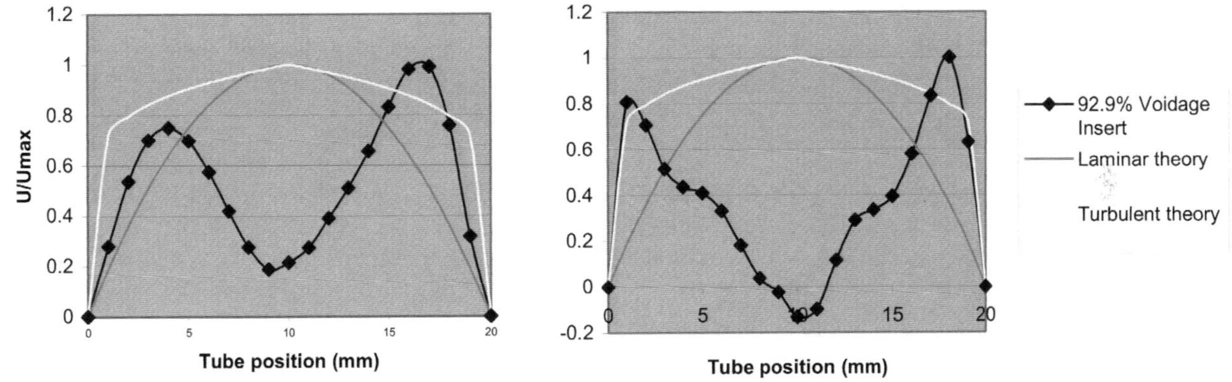

Figure 3 Velocity distribution across tube cross-section for 92.9% voidage element, compared to theoretical results for laminar and turbulent flow regimes. Reynolds number 100 on left and Reynolds number 500 on right

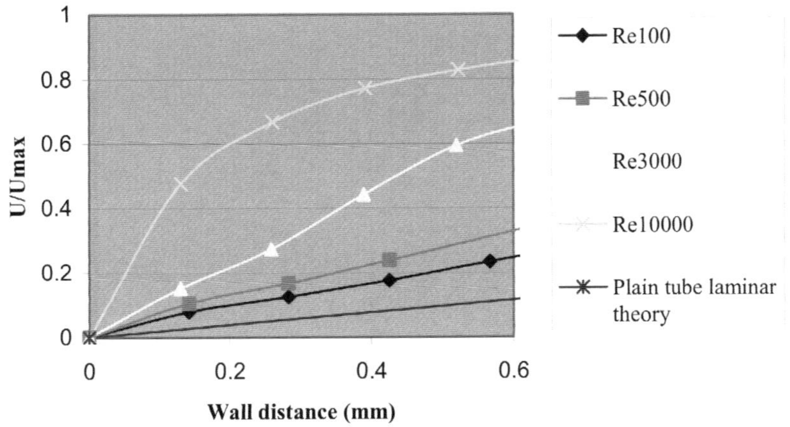

Figure 4 Velocity distributions close to tube wall for different Reynolds numbers. (97.3% voidage element)

How matrix elements work in practice.

An example is presented showing how an existing exchanger, used to heat a hydrocarbon by steam, can be modified to include hiTRAN® elements, resulting in an exchanger design with reduced fouling potential. hiTRAN.SP® software, developed to optimise tube-side enhancement, was used in conjunction with a commercially available exchanger design programme, to calculate the example.

The existing exchanger had been designed to use the available tube-side pressure drop to maximise heat transfer rates by utilising an 8-tube pass arrangement. The steam at 19 barg provides a good temperature driving force in a strongly tube side controlled scenario but this results in high tube-wall temperatures as illustrated in figure 5.

In order to reduce these tube-wall temperatures, whist maintaining the duty, the tube-side co-efficient must be substantially increased, leading to a reduction in steam temperature. In the hiTRAN® enhanced design, the exchanger is converted from 8 passes to 2 passes allowing the exchanger to operate within the original available pressure drop. The effect of these changes on the tube-wall temperature can be seen in figure 6. Lowering the wall temperature below the threshold value for fluid degradation (fluid cracking) would be beneficial in any service where this temperature dependant fouling mechanism exists.

In practice there are often more than one fouling mechanism acting simultaneously within an exchanger. Deposition (particle) fouling is often found in combination with degradation fouling. Both of these mechanisms are reduced by increased shear rates at the wall and enhanced radial mixing. Matrix elements provide these effects at low flow velocities, whist to obtain similar effects using plain tubes would require high flow velocities with associated high pressure loss and erosion potential.

This design modification is a low cost / risk option as the only changes necessary are the purchase and fitting of the elements and replacement or modification to the headers. Whilst this example indicates the options for an existing exchanger, the freedom available with a new unit can allow a designer to optimise the exchanger for reduced size and reduced fouling potential.

Table 1 Performance details of plain tube and hiTRAN® enhanced heat exchanger

		Plain tube	**hiTRAN®**
Shell type		AES	AES
Tube length	[m]	6	6
No of passes	[-]	8	2
Length of flow path	[m]	48	12
Performance details			
Temperature inlet/outlet	[°C]	120 / 165	120 / 165
Re inlet/outlet	[-]	1500 / 3500	370 / 870
Overall HTC	[W/m²K]	180	525
Tubeside HTC	[W/m²K]	243	935
Tubeside pressure drop	[kPa]	106	106
Tubeside flow velocity	[m/sec]	1.38	0.34
Tubeside residence time	[sec]	35	35
Shellside av. Walltemp	[°C]	211	172
Shellside steam pressure/Temp	[bar / °C]	20bar / 212°C	8.5bar / 172.5°C
Tubeside ave. Wall temp	[°C]	205	165
Tubeside ave. Bulk temp	[°C]	142	142
EMTD	[°C]	67	23

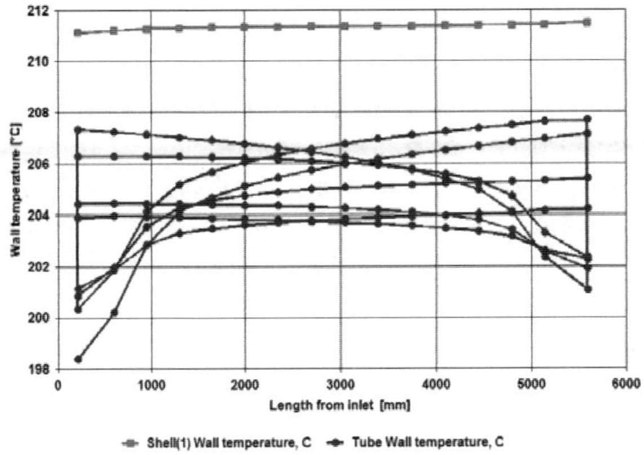

Figure 5 Plain tube wall temperatures

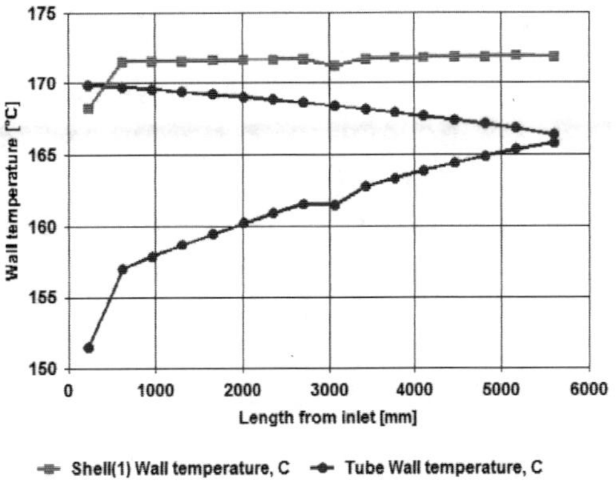

Figure 6 hiTRAN® wall temperatures

Shown in figure 7 is a screenshot of the programme used to calculate the heat transfer coefficient (hiTRAN.SP®) for tubes equipped with the wire matrix elements. The coloured curves showing the heat transfer coefficient in the enhanced tubes cover the full range of element void fractions available, starting with the highest void fraction (lowest density element) at the bottom of the curve and the lowest void fraction (highest density element) at the top of the curve.

When designing a heat exchanger the number of passes selected is usually based on the allowable pressure drop. In this case a two or three pass exchanger could be designed using different void fraction elements for the same (106kPa) pressure drop. A two-pass arrangement was favoured due to reduced wall temperatures and simplicity of modification of the existing 8-pass exchanger.

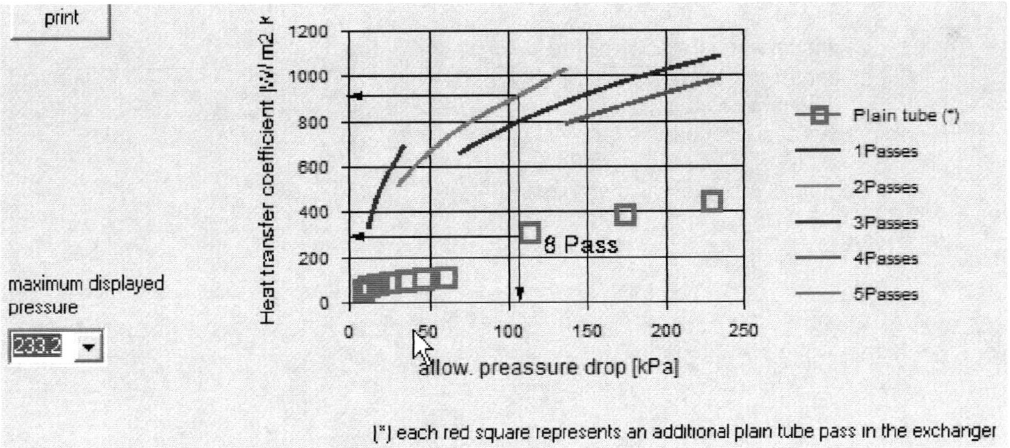

Figure 7 Screenshot of hiTRAN.SP® programme

Conclusions

Whilst it is fully acknowledged that there is unlikely to be a universal solution to preventing all aspects of hydrocarbon fouling by using a particular type or configuration of exchanger, it can be shown than in many cases reduced levels of fouling can be achieved with enhancement and in some cases the requirement for regular cleaning is actually eliminated. Where elevated or reduced temperature is the main precursor to degradation or crystallisation fouling respectively, then changing the wall temperature by means of enhancement can be shown to be highly beneficial. Where suspended particles settle on the tube wall due to local laminar conditions, creating turbulence provides a preventative function reducing actual deposition.

Practical application of these methods has been established where both of these fouling mechanisms were combining to limit operational life of 6 exchangers preheating a tar mixture, including distillation bottoms, in the manufacture of carbon black[5]. Installed in 1981/2 the unit now has 22 years of operation. Examining some of the elements removed from the exchangers over this period it was found that some amount of fouling does occur but has not been found to be a significant technical or commercial downside by comparison with the large benefits now relied upon.

Notwithstanding the successes gained, further work is required to meet the demand to find solutions to fouling problems, the identification of optimum operating velocity for different element void fractions in different services forms the highest priority. From such information, guidelines for designing heat exchangers that maintain performance can begin to be established. Research into preventing deposition fouling is currently being carried out at The University Of Edinburgh, UK.

Nomenclature

EMTD	effective mean temperature difference (K)
HTC	heat transfer coefficient (W/m^2K)
r	tube inner radius (m)
s	radial distance (m)
U_x	axial velocity in tube (m/s)
U_{mean}	mean velocity in axial direction (m/s)

Acknowledgements and References

The authors wish to thank the University of Birmingham Chemical Engineering Department for the use of their LDV equipment and their valuable expertise.

1 B Petty and A Karp, Optimal cleaning strategies in heat exchanger networks, Process Technology Quarterly, Autumn 2000 pp 111-117

2 N Epstein, Fouling in heat exchangers, In Heat Transfer 1978, Proc 6th Int. Heat Transfer Conference, Vol 6, Hemisphere (1979)

3 B.D Crittenden, S.T Kolaczkowski and T Takemoto, Use of in-tube inserts to reduce fouling in crude oils, AIChE Symp. Series, Vol 89, No 295, pp 300-307, 1993

4 J.M Coulson & J.F. Richardson, Chemical Engineering Volume 1, Pages 62-7, Pergamon Press, 1990

5 M.J Gough and J.V Rogers, Reduced fouling by enhanced heat transfer using wire matrix radial mixing elements, AIChE Symp Series 83 (No 257) pp 16-21 (1987)

Theoretical and Experimental Study of Fouling in Oil Reservoir and Production Equipment during Water Injection

J. Moghadasi[1,3], H. Müller-Steinhagen[2], , A. Sharif[3] and M. Jamialahmadi[1]

[1] Petroleum Engineering Department, Petroleum University, Ahwaz, IRAN.
E-mail: J.Moghadasi@surrey.ac.uk
[2] Institute for Thermodynamics and Thermal Engineering, Stuttgart University, D-70550, Stuttgart, GERMANY. E-mail: hms@itw.uni-stuttgart.de
[3] Chemical and Process Engineering Department, University of Surrey, Guildford, GU2 7XH, UK. E-mail: A.Sharif@surrey.ac.uk

Abstract

Due to the extensive use of water injection for oil displacement and pressure maintenance in oilfields, many reservoirs experience the problem of scale deposition and fouling when injection water starts to break through. In most cases the scaled-up wells are caused by the formation of sulfate and carbonate scales of calcium and strontium. Due to their relative hardness and low solubility, there are limited processes available for their removal and preventive measures such as the `squeeze' inhibitor treatment have to be taken. It is therefore important to gain a proper understanding of the kinetics of scale formation and its detrimental effects on formation damage under both inhibited and uninhibited conditions.

This paper presents the results of an experimental and theoretical study of permeability reduction of porous media caused by scaling. Two incompatible solutions of calcium and /carbonate ions were injected into the porous medium, where calcium or calcium carbonate was generated by chemical reaction. Mechanisms by which a precipitate reduces permeability include solids deposition on the pore walls due to attractive forces between the particles and the surfaces of the pores, individual particles blocking pore throats, and several particles bridging across a pore throat. The characteristics of the precipitate influence the extent of formation damage. Conditions such as large degree of supersaturation, presence of impurities, change in temperature, and rate of mixing controls the quantity and morphology of the precipitating crystals.

Introduction

The formation of mineral scale associated with the production of hydrocarbon has been a concern in oilfield operation. Depending on the nature of the scale and the fluid composition, the deposition can take place within the reservoir which causes formation damage [14, 17, 20, 25] or in the production facilities where blockage can cause severe operational problems. The two main types of scale which are commonly found in the oilfield are carbonate and sulphate scales [19-20, 27-30, 32, 36, 39-44]. Whilst the formation of carbonate scale [3, 42-43] is associated with the pressure and pH changes of the production fluid, the occurrence of sulphate scale is mainly due to the mixing of incompatible brines, [21-23, 25-26, 43-44] i.e. formation water and injection water. In the oilfield, the universal use of sea water injection as the primary oil recovery mechanism and for pressure maintenance means that problems with sulphate scale deposition, [25-26] mainly calcium and strontium, are likely to be present at some stage during the production life of the field. A field example is the Iranian offshore of Siri field [25] in the southern Persian Gulf. The oil in this field comes from a formation called Mishrif; this formation is common between Iran and the United Arab Emirates. Water injection into the Siri field was started in 1984 with 9100 bbl/day in order to maintain the pressure and to increase the oil recovery. But the injectivity decreased rapidly by 1990 the water injection was only 2200 bbl/day, and subsequently the water injection was stopped. The history of water injection in the Siri field is shown in Fig.1 [25] with a drop in injectivity of approximately 7000 bbl/day over a period of six years.

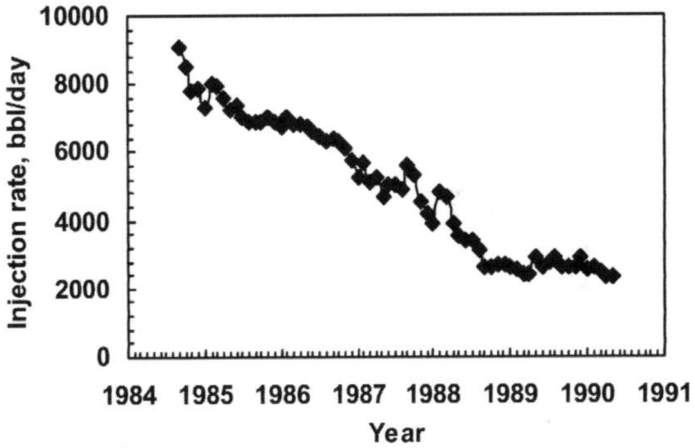

Fig. 1- Water injection history in Iranian Siri oil field

Calcium sulphate scale, especially gypsum, still causes unpredictable problems in many oil fields [1, 4, 31]. This type of scale can cause severe plugging of equipment and producing formations; it creates the necessity of costly remedial work. Thus, to understand the formation of these scale minerals, the kinetics of calcium sulphate scale formation and transformation of one hydrate to another are particularly important in determining the nature of the scale formed under field conditions as a function of both temperature and ionic strength of the solution.

1.1 Damage Mechanisms

Scale formation in surface and subsurface oil and gas production equipment has been recognised to be a major operational problem .It has been also recognised as major causes of formation damage either in injection or producing wells. Scale contributes to equipment wear and corrosion and flow restriction, thus resulting in a decrease in oil and gas production. Experience in the oil industry has indicated that many oil wells have suffered flow restriction because of and fouling within the oil –producing formation matrix and the downhole equipment, generally in primary, secondary and tertiary oil recovery operation as well as scale deposits in the surface production equipment.

Oil field scales costs are high because of drastic oil and gas production decline, frequently pulling of downhole equipment for replacement, reperforation of the producing intervals, reaming redrilling of the plugged oil wells, stimulation of the plugged oil-bearing formation, and other remedial workovers. As scale deposits around the wellbore, the porous media of formation becomes plugged and may be rendered impermeable to any fluids. Many case histories [18, 20, 26, 28-30, 36-37, 40, 42-48] of oil well scaling by calcium carbonate, calcium sulphate, strontium sulphate and barium sulphate have been reported. Problems pertaining to oil well scaling in North Sea fields have been reported [24] and are similar to cases in the Russia where scale has severely plugged wells. Oilfields scale problems have occurred as a result of water flooding in Algeria, Indonesia in south Sumatra oilfields, Saudi oil fields and Egypt in El-Morgan oilfield [12] where calcium and strontium sulphate scales have been found in surface and subsurface production equipment.

1.2 Scale problems in oil fields

Scale deposition and fouling is one of the most important and serious problems which water injection systems are generally engaged in. Scale sometimes limits or blocks oil and gas production by plugging the oil-producing formation matrix or fractures and perforated intervals. Scale also deposited in downhole pumps, tubing, casing flowlines, heater treaters, tanks, and other production equipment and facilities. Scale can occur at /or downstream of any point in the production system, at which supersaturation is generated. Supersaturation can be generated in single water by changing the pressure and temperature conditions or by mixing two incompatible waters. The most common oilfield scales deposited are calcium carbonate, calcium sulphate, strontium sulphate and barium sulphate. Calcium carbonate scale generally causes a sharp reduction in pressure such as that exists between the formation and the well bore and across any constriction in the production tubing, e.g. checks and safety valves. The reduction in pressure liberates CO_2 into the gas phase leaving a solution, which is supersaturated in calcium carbonate. The various forms of calcium sulphate scale i.e. gypsum, anhydrate and hemihydrate can be formed due to an increase in temperature. Fig. 2 gives some indication of which changes occur at which part of an oilfield.

1.3 Type of scales in oil fields

The most common oil field scales are listed in Table 1, along with the primary variables, which affect their solubility.

Table 1
Most Common Oilfield Scales

Name	Chemical Formula	Primary Variables
Calcium carbonate	$CaCO_3$	Partial pressure of CO_2 ,Temperature, Total dissolved salts, pH
Calcium Sulphate:		
Gypsum	$CaSO_4.2H_2O$	Temperature, Total dissolved salts, Pressure
Hemihydrate	$CaSO_4.1/2H_2O$	
Anhydrite	$CaSO_4$	
Barium Sulphate	$BaSO_4$	Temperature, Pressure
Strontium Sulphate	$SrSO_4$	Temperature, Pressure, Total Dissolved salts

1.4 The nature of calcium sulphate scale

1.4.1 ***Composition***

Calcium sulphate is a crystalline deposit that is very adherent to many surfaces. It is composed mainly of calcium and sulphate ions, but when deposited from complex polymetallic solutions can contain traces of many other ions. Calcium sulphate often co-precipitates with strontium sulphate, with which it can form a solid solution. Additionally on precipitation from oilfield fluids, it may contain small amounts of wax, rust and silt.

In downhole conditions either anhydrite or gypsum deposition is caused mainly by pressure drop, which has a bigger effect than temperature. Harberg et al. (1992) investigated the effect of brine ion concentration, temperature and pressure on gypsum precipitation.

1.4.2 Morphology of Calcium Sulphate

Calcium sulphate, which is important in desalination geochemistry and petroleum engineering, is complicate by the fact that it can crystallise from aqueous solution in three forms: gypsum ($CaSO_4$, $2H_2O$), hemihydrate ($CaSO_4$, $½H_2O$), and anhydrate ($CaSO_4$). These compounds may be stable depending on temperature and ionic strength and they have decreasing solubilities with increasing temperatures above 40°C. Nancollas et al. (1978) and many investigators (Dickson et al., 1963, Vetter et al., 1983) have studied the morphology of scale.

1.4.3 Calcium Sulphate Solubility

Solubility is defined as the limiting amount of a solute, which can be dissolved in a solvent under a given set of physical conditions. The chemical species of interest to us are present in aqueous solutions as ions. Certain combinations of these ions lead to compounds, which have low solubility. Once this capacity or solubility is exceeded the compounds precipitate from solution as solids. Therefore, precipitation of solid materials, which may form scale, will occur if:

- The water contains ions, which are capable of forming compounds of limited solubility.

- There is a change in the physical conditions or water composition, lowering the solubility.

Factors that affect scale precipitation, deposition and crystal growth can be summarised as: Supersaturation, temperature, pressure, ionic strength, evaporation, agitation, contact time and pH. Effective scale control should be one of the primary objectives of any efficient water injection and normal production operation in oil and gas fields.

Effect of temperature and pressure - Landolt-Bornstien (1985) shows the effect of temperature on solubility of calcium sulphate. Gypsum solubility increases with temperature up to about 40 °C, and then decreases with temperature. Note that above about 40°C, anhydrite becomes less soluble than gypsum, so it could reasonably be expected that anhydrite might be the preferred from of calcium sulphate in deeper, hotter wells. Actually, the temperature at which the scale changes from gypsum to anhydrite or hemihydrate is a function of many factors, including pressure, dissolved solids concentration, flow conditions, and the speed at which different forms of calcium sulphate can precipitate out from solution. Prediction which form of calcium sulphate will precipitate under a given set of conditions is very difficult. Even though an anhydrite precipitate might be expected above 40°C in preference to gypsum due to its lower solubility, gypsum may be found at temperature up to 100°C. It is often difficult to precipitate anhydrite directly from solution, but, with the passage of time, gypsum can dehydrate to form anhydrite. Above 100°C, anhydrite will precipitate out directly in a stirred or flowing system. If the system is quiescent the hemihydrate solubility becomes limiting. Conversion of hemihydrate to anhydrite could be expected to occur with time.

Dickson et al., 1963 shows the effect of pressure and temperature on anhydrite solubility. The solubility of calcium sulphate in water increases with pressure. This increase in solubility is due to the fact that when the scale is dissolved in water, there is decrease in the total volume of the system.

$$CaSO_4 + H_2O \rightleftharpoons Ca^{+2} + SO_4^{-2} + H_2O \qquad (1)$$

Pressure drop can be a major cause of calcium sulphate scale in producing wells and near the wellbore can create scale back in the formation as well as in the piping.

Effect of ionic strength - Ionic strength is defined as:

$$I = \sum C_i Z_i^2 \qquad (2)$$

The solubility of calcium sulphate is strongly affected by the presence and concentration of other ions in the system. The effect of ionic strength is shown in Fig. 3 (Carlberg and Matches, 1973).

Effect of pH on crystal growth - Schierholtz (1974) investigated the unseeded crystallisation of calcium sulphate dihydrate and followed the change in calcium concentration during the initial induction periods and subsequent growth at pH values ranging from 4.5 to 6.6. Within this range, the pseudo first order rate constant decreased by a factor of three. It was concluded that nucleation of calcium sulphate dihydrate is affected by pH. Austin et al. (1975) studied the spontaneous precipitation of calcium sulphate phase from simulated seawater in the temperature range of 125 to 150°C at pH 2.3 to 8.1. Under this condition, the hemihydrate phase precipitated first and there was a relatively slow transformation to the anhydrite phase with the rate increasing with temperature.

Effects of agitation and vaporization- Agitation and vaporisation can increase scale production. The evaporation of water evolves gas in or around the well and may cause supersaturated conditions to develop producing calcium sulphate.

1.5 The nature of calcium carbonate scale

Calcium carbonate or calcite scale is frequently encountered in oil field operations. Aragonite and Vaterite are the other crystalline forms of calcium carbonate, but the calcite has the greatest stability in oil field circumstances, so it is the most common form of calcium carbonate encountered in oil field production operation. Calcium carbonate crystals are large, but when the scale is found together with impurities in the form of finely divided crystals, then the scale appears uniform. Deposition of $CaCO_3$ scale or sludge results from precipitation of calcium carbonate according to the following equation:

$$Ca^{+2} + CO_3^{-2} \rightarrow CaCO_3 \qquad (3)$$

As it will be seen later calcium carbonate scale can also be formed by combination of calcium and bicarbonate ions, and this reaction is the major cause of calcium carbonate scale deposition and fouling in oilfield operations. This is because only a small percentage of the bicarbonate ions dissociate at the pH values found in most injection waters to form H^+ and CO_3^{-2}, as shown in Fig. 4. (Patton, 1968)

1.5.1 Calcium carbonate solubility

Carbon dioxide partial pressure effect – As opposed to most sulphate scales the prediction of carbonate scales requires not only the consideration of pressures, temperatures and water composition, but also the knowledge on the chemical reactions within the brine and CO_2 in the gas phase. Most oilfield reservoirs contain carbonate mineral cements and carbon dioxide; therefore the formation water is normally saturated with calcium carbonate under reservoir conditions where the temperature can be as high as 200°C and the pressure up to 30 MPa. When carbon dioxide comes in contact with water, it dissolves and forms carbonic acid according to equation (4). The ionisation of carbonic acid is illustrated by the following equations:

$$CO_2 + H_2O \leftrightarrow H_2CO_3 \qquad (4)$$

$$H_2CO_3 \leftrightarrow H^+ + HCO_3^- \qquad (5)$$

$$HCO_3^- \leftrightarrow CO_3^{-2} + H^+ \qquad (6)$$

From the above equations, it can be seen that carbon dioxide gas dissolves in water and forms carbonic acid. The carbonic acid ionises to form hydrogen ion and bicarbonate ion. Since the second ionisation constant of carbonic acid is much smaller than the first ionisation constant, bicarbonate ions vastly outnumber the number of carbonate ions present under normal circumstances. It is believed that dissolved calcium carbonate does not exist in solution as calcium ions and carbonate ions, but as calcium ions and bicarbonate ions.

Thus, the precipitation of calcium carbonate can be expressed by the following equation:

$$Ca^{+2} + 2HCO_3^- \leftrightarrow Ca(HCO_3)_2 \leftrightarrow H_2O + CO_2 + CaCO_3 \downarrow \quad (7)$$

By the principle of LeChatelier, we see that by increasing the concentration of carbon dioxide, more calcium bicarbonate is formed. A decrease in carbon dioxide content in this system at equilibrium would result in the formation of calcium carbonate. Therefore, it can be seen that the solubility of calcium carbonate is greatly influenced by the carbon dioxide content of the water. The amount of CO_2 that will dissolve in water is proportional to the partial pressure of CO_2 in the gas over the water, which is the product of mole fraction of CO_2 in gas phase and the total pressure of the system. So, if either the system pressure or the percentage of CO_2 in the gas were to increase, the amount of CO_2 dissolved in the water would also increase. Figs. 5 and 6 illustrates the effect of CO_2 partial pressure on the pH of the water containing little or no dissolved minerals and on solubility of $CaCO_3$ in pure water. This data illustrates that $CaCO_3$ solubility increases with increased CO_2 partial pressure. The effect becomes less pronounced as the temperature increases. The reverse is also true. It is one of the major causes of $CaCO_3$ scale deposition. At any point in the system where a pressure drop is taken, the partial pressure of CO_2 in the gas phase decreases, CO_2 comes out of solution, and the pH of the water rises. This shifts reaction equation (7) to the right and may cause $CaCO_3$ precipitation.

Effect of total pressure: The solubility of calcium carbonate in a two-phase system increases with increased pressure for two reasons:

- Increased pressure increases the partial pressure of CO_2 and increases the solubility of $CaCO_3$ in water as previously explained.
- Increased pressure also increases the solubility due to thermodynamic considerations, as has been discussed for the case of calcium sulphate.

Effect of pH - The amount of CO_2 present in the water affects the pH of the water and the solubility of calcium carbonate. However it really does not matter what causes the acidity or alkalinity of the water. The lower the pH, the less likely is $CaCO_3$ precipitation. Conversely the higher the pH, the more likely that precipitation will occur.

Effect of temperature- Contrary to the behavior of most materials, calcium carbonate becomes less soluble as temperature increases. The hotter water the more likely the $CaCO_3$ precipitation. Hence, water, which is nonscaling at the surface, may result in scale formation in the injection well if the downhole temperature is sufficiently high. Plummer and Busenberg (1982) show the general behavior of $CaCO_3$ solubility as a function of temperature.

Effect of dissolved salts - Calcium carbonate solubility increases as the dissolved salt content of the water increases. Actually, the higher the total dissolved solids (not counting calcium or carbonate ions), the greater the solubility of $CaCO_3$ in the water and the lower the scaling tendency up to a maximum of about 200 g/l.

Previous Work

The formation of precipitate starts with nucleation processes in which attracted atom join to form submicron nuclei [1, 7-8]. When impurities are present in the fluid, the energy required to form these nuclei is smaller than in a pure solution. The impurities act as nucleation sites, and this process, called heterogeneous nucleation, occurs commonly in subsurface reservoirs where fluids contain mineral particles in suspension.

Experimental Studies- several experiments studied the extent of permeability damage caused by flowing precipitates in cores and sand packs. Some investigators tested incompatible waters from the North Sea [24, 39], Middle East and El-Margon [12] oilfield that produced barium, calcium and strontium sulfate precipitates. They reported that 15 to 20 μm crystals blocked the pore throats inside glass-bead packs and alumina cores by size exclusion and bridging mechanisms. Crystals often grew perpendicular to the pore walls, and some crystal aggregates also had the perpendicular to the pore walls, and some crystal aggregates also had the form of "books" and rosettes. Todd and Yuan [39] also conducted laboratory investigations using North Sea reservoir brines that produced barium and strontium sulfate scales. Crystals depositing along and growing perpendicular to the pore surface caused most of the reduction in core permeability. They observed that doubling the supersaturation ratio of both barium and strontium sulfate produced an increase in the quantity of scale formed inside the pores and a change in the morphology of the crystals. Both changes increase the rate of permeability decline. The basic structures of crystal formed under dynamic flow conditions and in a static solution are similar, but some faces of the crystal disappear and others become more plate-like when subjected to shear forces. Injection tests on cemented glass-bead packs showed a 95.5 % reduction in permeability because of the bridging of 2 to 10 μm calcium carbonate crystals that blocked 33 μm pore throats [9]. Thronton and Lorenz [38] and Crow [8] showed how sodium and iron hydroxide colloid particles block the pores of quartz sand. They suggested two mechanisms by which the plugging of quartz sand occurs: flocculation/ coagulation of the sodium and iron hydroxide phase producing a filter cake near the inlet of the sand packs and colloid-quartz surface interaction by which the particles attach to the quartz surface.

Theoretical Studies- geochemical flow models were developed [6, 14-16, 32, 45-46, 48] for both mineral dissolution and precipitation processes during convective transport. These models were based on the assumptions of local thermodynamic equilibrium and negligible dispersion. These models did not allow for solid migration, assumptions that were severely limiting because experimental studies demonstrate that precipitated solids can migrate with the fluid. A theory of solids fractional flow was proposed in an attempt to characterize the local permeability reduction by incorporating the dominant pore-plugging mechanism [47]. A fractional flow function for each migrating mineral, coupled with the mass conservation of each element and the assumption of local equilibrium between the solid and aqueous species, provided a complete description of reaction during flow through permeable media.

Experimental setup and procedure

Experiments were carried out using a test rig, which is schematically shown in Fig. 7. The stainless steel tank is heated to approximately 50°C using four band heaters mounted on the outside. A small cooling water coil is installed inside the tank to allow accurate control of the liquid temperature in the tank. The thermocouple for determining the temperature inside the tank is connected to a controller for the band heaters power output. Furthermore, a stirrer is also placed inside the tank to agitate the liquid and to provide a uniform temperature. The peristaltic pump speed can be varied between 0.5-55 rpm so that the liquid flow can be accurately adjusted. The maximum design pressure of the pump is 3 bars, which is achieved by using tubing with an internal diameter of 1.6 mm and a wall thickness of 1.6 mm. The tubing is made from Marprene II, a material that is resistant to water and mineral oil. For the scale formation experiments, a second, identical set of tank and peristaltic pump has been installed to provide the scale forming solution mixture, which was drained after passing through the test section.

The test section, which contains the porous medium, is made of stainless steel pipe with an internal diameter of 32 mm, a wall thickness of 5 mm and a total length of 580 mm. Spiral and longitudinal grooves to accommodate heating wire and thermocouples for measuring the wall temperature have been milled in the outside of the pipe. Bores to insert thermocouples for measuring the sand bed temperature, and opposite to them bores for pressure tapping, have been drilled at locations shown in Fig. 8. The cold-ended thermocoax heating wire is fitted inside hemispherical grooves around the tube at a pitch of 8 mm, to give a heated area of 0.03488 m^2. The power (maximum 2000 W) is controlled to maintain the wall temperature below the maximum operating temperature of around 200 °C. The test section has six pressure taps along its length, each of tem connected to a separate pressure transducer. An additional perspex test section has been manufactured and installed for visual observation of the deposition mechanisms.

All transducers have been connected to a computer-controlled data acquisition system for on-line monitoring and processing of the experimental results.

The packed bed is a direct method for studying deposition mechanisms from particle suspensions and salt solutions. In particle deposition with the packed bed technique, a well defined granular material which is usually made of spherical glass beads or washes sand is packed in a column to form a porous bed with a fixed porosity. When the bed is fully packed, the porous medium fills the space between the two screens at the ends of the test section. The pore volume of the dry porous medium was then filled with liquid supplied from a burette to determine the porosity as the ratio of the required volume of liquid divided by the total volume of the bed. The same procedure was repeated several times for each medium and the mean value was taken to represent the porosity of the medium. The properties of the packing materials and of the investigated fluids are given in Table 2.

Preliminary tests were performed to obtain the time after which the bed was stabilized: Distilled water was pumped through the bed for about one hour to obtain a homogenous condition. While all operational variables of the system were kept constant, pressure readings were taken at short time intervals until steady state was reached. This was usually achieved after about 20 minutes. The scale formation experiments were performed with aqueous solutions of Na_2SO_4, $Ca(NO_3)_2.4H_2O$ and Na_2CO_3. The criteria for selecting these salts were based on the solubility of the salts and the valence of the respective ions. Each test liquid was mixed from two solutions of these salts, one rich in calcium ions and the other rich in sulphate or carbonate ions, which were kept separate until entering the porous medium. Mixing of calcium nitrate ($Ca(NO_3)_2.4H_2O$) and sodium sulphate (Na_2SO_4) or sodium carbonate results in the formation of calcium sulphate or calcium carbonate in the porous medium. The range of salt concentrations used in this investigation is given in Table 3. Calcium nitrate and sodium sulphate or sodium carbonate solutions of predetermined concentrations were filled into the supply tanks. Distilled water was fed through the bed until the system reached a steady state at the desired temperature and flow velocity. Then the data acquisition system was switched on to record fluid flow rate, temperatures, and pressures. At this time the supply pumps were switched to the scale forming solutions for about 1600-1800 minutes then the supply pumps were switched to EDTA (ethylene diamine tetra acetic acid) solution and flooding continued for about 180-400 minutes. The range of investigated operating variables is given in Table 3.

Table 2 Physical properties of packing material

Packing material	Particle size range, μm	Mean particle diameter, μm	Porosity	Permeability, m^2
Sand	180-250	192	0.3827	2.14E-11
	250-425	265	0.3830	6.03E-11
	400-500	410	0.3839	1.21E-10
	1000	1000	0.3853	7.27E-10
Glass	180-300	245	0.3787	4.41E-11
	250-425	338	0.3792	7.89E-11
	400-600	480	0.3804	1.59E-10
	1000	1000	0.3825	7.05E-10

Table 3 Range of operating parameters in scale formation experiments

Inlet temperature	50°C-80°C	
Flow rates	25cm^3/min -100cm^3/min	
System pressure	122 kPa	
Solution viscosity	0.7×10^{-3}-1.3×10^{-3} kg/m.s	
Solution concentration	$Ca(NO_3)_2.4H_2O$	1.8 - 13.4 g/l
	Na_2SO_4	2.6 - 7.8 g/l
	Na_2CO_3	0.8 - 1.6 g/l
	$CaSO_4$	2.5 - 7.5 g/l
	$CaCO_3$	0.75 - 1.5 g/l

3. Results and discussion

3.1 Clean bed experiments

Before carrying out any tests with solutions, it is first necessary to have adequate information regarding the flow mechanisms in a clean medium. When a fluid flows through a porous medium, the pressure drop, which develops along the bed in the direction of flow, is a function of system geometry, bed voidage and the physical properties of bed and fluid. The operating conditions can result in four distinct flow regimes [11]: Darcy or creeping flow, inertial flow, unsteady laminar flow and chaotic (or turbulent) flow. In the Darcian region the pressure gradient is proportional to the flow rate and is mathematically expressed by

$$u = \frac{K}{\mu}\left(-\frac{\Delta p}{\Delta x}\right) \tag{10}$$

The coefficient K for single-phase flow depends only on the geometry of the porous medium. It is called the specific or absolute permeability of the medium; in the case of single-phase flow, this is abbreviated as permeability. The measurements of pressure drop as a function of axial distance are presented in Fig. 9 for different media and liquid flow rates. As predicted by eq. (10), there is a linear relationship between the pressure drop and the axial distance in the direction of flow.

Knowing the viscosity of the liquid saturating the medium, the permeability of each medium can be calculated using the slope of the best-fit straight line through the data presented in Fig. 10 as the ratio of $\mu u/K$ in equation (10). The calculated permeabilities are also given in Table 2.

3.2 Scale formation experiments

The main objective of this part of the investigation is to study permeability reduction caused by calcium sulphate and calcium carbonate scale deposition and fouling in porous media. A wide range of flow velocities, bulk temperatures and fluid bulk concentrations are considered. During each run the pressure drop across the test section was recorded continuously. The pressure drop increased during the experiments only when a supersaturated solution was flowing through the test section. This confirms that the increase is caused by scale formation. The change of permeability and the pattern it follows are the most significant pieces of information to be gained from the experimental study. In the Figs. 11-16, the results for various flow rates, temperatures and concentrations are drawed individually.

At higher flow rates more calcium, carbonate and sulphate ions will enter the porous medium in a given interval of time, hence providing more material for deposition (see Figs. 11-12). The permeability decline is more rapid at higher temperature, since the rate of precipitation and the supersaturation both increase with temperature (see Figs. 13-14).

Figs. 15-16 show the variation in permeability decline with time for different concentrations. When the concentration of the solution (i.e. supersaturation) is increasing, plugging and hence permeability loss occurs more rapidly.

Conclusions

1. Permeability decline caused by scale formation in the porous bed ranged from less than 30% to more than 90% of the initial permeability, depending on the solution composition, initial permeability, temperature, flow rate and solution injection period. The pattern of permeability decline in a porous medium due to continues scaling was characterized by a concave curve with a steep initial decline, which gradually decreased to a low, but often significant, constant damage rate. The initial steepness of these curves generally decreased with increasing distance from the point of mixing of the incompatible solutions. The shape of the curve was related to changes in the pattern of growth of the scale crystals. The concave shape of the permeability-time curves was common to the majority of the porous medium flow tests.
2. Many influencing factors on scale formation have been examined. Temperature change had a remarkable effect on the scaling rate. At higher temperatures calcium sulphate and calcium carbonate scale is increased because the solubilities of calcium sulphate and calcium carbonate decrease with temperature. This must have increased the rate of precipitation and consequently the permeability decline.
3. It was observed that brines of higher degree of supersaturation produced a more rapid decline in permeability. This was to be expected since increased supersaturation would result in a more rapid rate of scale precipitation.
4. As the flow rate was increased, the rate of permeability decline becomes more rapid. At higher flow rates more calcium and sulphate/carbonate ions will pass through the porous medium in a given interval of time. The supersaturation will therefore increase, producing a higher rate of precipitation. This increased precipitation rate will produce a larger overall permeability decline. Flow rate causes more for calcium carbonate precipitation than calcium sulphate.
5. The modelling performed in this work was of a preliminary nature to investigate the suitability of the chemical kinetic modelling technique for simulation of scale formation in a porous medium. The model simulated, reasonably well, the pattern of permeability decline observed in the experimental work. Permeability decline curves predicted by the model were of a similar shape to those produced experimentally. The pattern of increased initial damage rates with higher initial permeabilities evident in the experimental work was also apparent in the results from the model.
6. Permeability decline predicted by the model increased with increasing flow rate, temperature and concentration of the solutions. Quantitatively, the magnitude of the permeability change was similar to that demonstrated in the experimental work.

Nomenclature

A	cross-sectional area, m^2
C	total amount of ionic species in solution, M
C_s	mass solid concentration, kg/m^3
d_c	test section diameter, m
d_p	particle diameter, □m
dp	pressure drop, N/m^2
dp/dx	pressure gradient in x direction, N/m^3
EDTA	ethylene diamine tetra acetic acid
I	Ionic strength, mole
K	permeability, m^2
m	mass fraction, slope or rate of scale deposited, kg/m^3
p	pressure, N/m^2
$\Delta p/l$	pressure gradient, Pa/m
Δp	pressure drop, Pa
Q_0	cumulative volume of fluid injected, m^3/s
Q	volumetric flow rate, m^3/s
Re_m	modified Reynolds number, $d_p \cdot \rho \cdot u / \mu \cdot (1-\phi)$
Re_f	Reynolds number defined as $d_p \cdot \rho \cdot u_f/\mu$
T	temperature, °C, K°
t	time, min, s
u	velocity of flowing phase through porous medium, m/s
x	distance from inlet face of test section or core, m

Greek Symbols

ϕ	Porosity
μ_f	fluid viscosity, kg/m.s
∇	gradient

Subscripts

hr	hour
i	initial or inlet
l	liquid
L	length
PV	pore volume
PVI	pore volume injection, permeability variation index
T	tube, pore
TC	thermocouple

References

1. Atkinson, G., Raju, K. and Howell R.W.: "The Thermodynamics of Scale Prediction," paper NO. 21021, present at the Society of Petroleum Engineers International Symposium on Oilfield Chemistry, Anaheim, CA, 1991.
2. Austin, A.E., Miller, J.F. et al.: "Precipitation of Calcium Sulphate From Sea Water at High Temperature," Desalination, Vol. 16, pp.331-339, 1975.
3. Bezmer, C. and Bauer, K.A.: "Prevention of Carbonate Scale Deposition- A well Packing Technique with Controlled Solubility Phosphates," J. Pet. Tech., Vol. 246, pp. 505-14, 1969.
4. Blout, C.W. and Dickson, F.W.: "Gypsum-Anhydrite Equilibrium in Systems $CaSO_4$ and $CaSO_4$-NaCl-H $_2$O," Amer. Min, Vol. 58, pp. 1746, 1973.
5. Carlberg, B.L and Mathews, R.R.: "Solubility of Calcium Sulfate in Brine," paper No. 4353 presented at the Society of Petroleum Engineers Oilfield Chemistry Symposium, May 24-25, 1973.
6. Civan, F., Kapp Roy, M. Ohen Henry, A.: "Alteration of Permeability of Fine Particle Processes," Journal of Petroleum Science and Engineering, No. 3, pp. 65-79, 1989.
7. Cowan, J.C. and Weintritt, D.J.: "Water Formed Scale Deposits," Gulf Publishing Co., pp. 586, Houston 1976.
8. Crow, C.W.: "Evaluation of Agents for Preventing Precipitation of Ferric Hydroxide From Spent Treating Acid," JPT, pp. 691-695, Apr. 1985.
9. Cusack, F., Mc-Kinley, V.L., Lappin-Scott, H.M., Brown, D.R., Clementz, D.M., and Costerton, J.W.: "Diagnosis and Removal of Microbial/Fines Plugging in Water Injection Wells," Paper SPE 16907 Presented at the 62nd Annual Technical Conference and Exhibition of the Society of Petroleum Engineers held in Dallas, TX September 27-30, 1987.
10. Dickson, F.W., Blount, C.W. and Tunell, G.: "Use of Hydro-Thermal Solution Equipment to Determine the Solubility of Anhydrite in Waters from 100 °C to 275 °C and from 1 bar to 1000 bars Pressure," American Journal of Science, Vol. 261, 61, 1963.
11. Dybbs, A. and Edwards, R. V.: "Department of Fluid, Thermal and Aerospace Science report," FTAS/Tr (Case Western Reserve University) No. 75-117, Workshop on Heat and Mass Transfer in Porous Media (PB-252-387) (NTIS: Springfield, VA) pp. 228, 197
12. El-Hattab, M.I.: "Gupco's Experiences with Treating Gulf of Squeeze Seawater for Water Flooding the El-Morgan Oil Field," J. Pet Tech., pp. 1449-60, July 1982.
13. Harberg, T., Selm I. et al.: "Scale Formation in Reservoir and Production Equipment During Oil Recovery: An Equilibrium Model," Society of Petroleum Engineer production Engineering, pp.75-84, Feb. 1992.
14. Khatib, Z.I.: "Prediction of Formation Damage Due to Suspended Solids: Modelling Approach of Filter Cake Build-up in Injectors," Society of Petroleum Engineers paper No. 28488, Society of Petroleum Engineer 69th Annual Technical Conference and Exhibition, New Orleans, La., Sep. 25-28, 1994.
15. Khillar, K.C. and Fogler, H.C.: "Colloidally Induced Fines Migration in Porous Media," Reviews in Chemical Engineering, Vol. 4(1 & 2), pp. 41-108, 1987.
16. Khillar, K.C. and Fogler, H.C.: "Colloidally and Hydrodynamically Induced Fines Migration in Porous Media," Chapter-19, in Encyclopaedia of Fluid Mechanics, Vol. 10, Surface and Ground Water Flow Phenomena, Ed. N.P. Cheremisinoff, pp. 623-663, 1990
17. Krueger, R.F.: "An Overview of Formation Damage and Well Productivity in Oil Field Operation", JPT, pp. 131-152, Feb. 1986.
18. Krumine, P.H., Mayer, E.H. and Brock, G.E.: "Scale Formation During Alkaline Flooding," JPT, pp. 1466-1474, Aug. 1985.
19. Landolt-Bornstien: "Zahlenwerte und Funktionen," Vol. 2, Sec. B, Springer-Vertag, Berlin, 1985.
20. Lasetr, R.M., Grander, T.R., and Glasscock, F.M.: "Scale Deposits are Controlled Now With Liquid Inhibitors," Oil

and Gas J., pp. 88-63, Jan. 15, 1968.
21. Leon, J.A. and Scott, E.M.: "Characterization and Control of Formation Damage During Water Flooding at High Clay content Reservoir," Paper SPE 16234 Presented at the SPE Production Operation Symposium, Oklahoma city, Mar. 8-10, 1987.
22. Lindlof, J.C. and Stoffer, K.G.: "A Case Study of Seawater Injection Incompatibility," J. Pet Tech., pp. 1256-62, July 1983.
23. Liu, S.T. and Nancollas, G.H.: "The Kinetics of Dissolution of Calcium Sulphate Dihydrate," Journal of Inorg. Nucl. Chem., Vol. 33, pp. 2311, 1971.
24. Mitchell, R.W., Grist, D.M., and Boyle, M.T.: "Chemical Treatments Associated with North Sea Projects," Journal of Petroleum Technology, pp. 904-12, 1980.
25. Moghadasi, J., Jamialahmadi, M., Müller-Steinhagen, H., Sharif, A., Ghalambor, A., Izadpanah, M.R. and Motaie, E.: "Scale Formation in Iranian Oil Reservoir and Production Equipment during Water Injection," Paper SPE 80406 Presented at the 5th International Oilfield Scale Symposium and Exhibition held in Aberdeen, UK, 29–30 January 2003.
26. Moghadasi, J., Jamialahmadi, M., Müller-Steinhagen, H., Sharif, A. and Izadpanah, M.R.: "Formation Damage in Iranian Oilfield," Society of Petroleum Engineers paper No. 73781, Presented at Society of Petroleum Engineer Formation Damage Control International Symposium and Exhibition, Lafayette, Louisiana, USA, 20-21 Feb., 2002.
27. Nancollas, G.H. and Campbell, J.R.: "The Crystallization and Dissolution of Strontium Sulphate in Aqueous Solution," Journal of Physical Chemistry, Vol. 23, No.6, pp. 1735-40, 1969.
28. Nancollas, G.H, Eralp, A.E and Gill, J.S.: "Kinetics of Growth of Calcium Sulphate Scale Formation: A Kinetic Approach," Society Petroleum Engineers Journal, pp. 133-138, 1978.
29. Nancollas, G.H. and Liu, S.T.: "Crystal Growth and Dissolution of Barium Sulphate," Society Petroleum Engineers Journal, pp. 509-16, 1975.
30. Nancollas, G.H. and Sawada, K.: "Formation of Scale Calcium Carbonate Polymorphs: the Influence of Magnesium Ion and Inhibitor," Journal of Petroleum Technology, pp. 645-52, March 1982.
31. Oddo, J.E and Tomson, M.B: "Why Scale Forms and How To Predict It," Society of Petroleum Engineers Production and Facilities, Feb. 1997.
32. Pang R. and Sharma, M.M.: "A Model for Predicting Injectivity Decline in Water Injection Wells," Society of Petroleum Engineers paper No. 28489, Society of Petroleum Engineers 69th Annual Conference and Exhibition, New Orleans, La., Sept. 25-28, 1994.
33. Patton, C.C.: "Applied Water Technology," Campell Petroleum Series, Norman, Oklahoma, 1968.
34. Plummer, L.N. and Busenberg, E.: "The Solubilities of Calcite, Aragonite and Vaterite in CO_2-H_2O Solutions Between 0 and 90°C and an Evaluation of the Aqueous Model for the System $CaCO_3$-CO_2-H_2O,"Geochimica et Cosmochimica Acta, Vol. 46, pp. 1011-1040, 1982.
35. Schierholtz, O.J.: "The Crystallisation of Calcium Sulphate Dihydrate," Cd. J. Chem., Vol. 36, pp. 1057-63, 1950.
36. Shen, J. and Corsby, C.C.: "Insight Into Strontium and Calcium Sulphate Scaling Mechanisms in Oil Producing Wells," J. Pet Tech., pp. 1249-55, July 1983.
37. Singley, J.E.: "Corrosion Prevention and Control in Water Treatment and Supply Systems," Noyes Publication, Houston, 1985.
38. Thronton, S.D. and Lorenz, P.B.: "Role of Silicate and Aluminate Ions in the Reaction of Sodium Hydroxide with Reservoir Minerals", Paper SPE 16277 Presented at the SPE International Symposium on Oilfield Chemistry, San Antonio, TX, Feb. 4-6, 1987.
39. Todd, A.C. and Yuan, M.D.: "Prediction of Sulphate Scaling Tendency in Oilfield Operations," SPE Production Engineering Journal, pp. 63-72, Feb. 1991.
40. Valone, W.F. and Skillern, K.R.: "An Improved Technique for Predicting the Severity of Calcium Carbonate," Proc. Society of Petroleum Engineers of AIME International Oil Field and Geothermal Chem. Symp., Dallas, Paper No. 10549, pp. 11-32, 1982.
41. Vetter, O.J.: "Oilfield Scale — Can We Handle It?," Journal of Petroleum Technology, pp.1402-08, Dec. 1972.
42. Vetter, O.J. and Farone, W.A.: "Calcium Carbonate Scale in Oilfield Operations," Society of Petroleum Engineers paper No. 16908 presented at the 62nd Annual Technical Conference and Exhibition held in Dallas, September 27-30, 1987.
43. Vetter, O.J., Farone, W.A., Veith, E. and Lankford, S.: "Calcium Carbonate Scale Considerations: A Practical Approach," Society of Petroleum Engineers paper No. 17009 presented in the International Symposium of Production Technology held in Lubbock, November 16-17, 1987.
44. Vetter, O.J., Kandarpa, V. and Harvaka, A.: "Predication of the Scale Problems Due to Injection of Incompatible Waters", Journal of Petroleum Technology, pp.273-284, Feb. 1982.
45. Vitthal, S. and Sharma, M.M.: "Dynamics Model for Particle Deposition and Build up in Granular Media," Journal of Colloidal and Interface Science, 153, No. 2, pp. 314-336, Oct. 1992.
46. Walsh, M.P., Rouse, B.A., Senol, N.N., Pope, G.A., and Lake, L.W.: "Chemical Interactions of Aluminum-Citrate Solutions with Formation Minerals," Paper SPE 11799 Presented at the International Symposium on Oilfield and Geothermal Chemistry held in Denver, CO, June 1-3, 1983.
47. Wu, G.G.: "A PC-Windows-Based Program for Geostatistical Modelling Application," SPE Paper 28242 Presented at the SPE Petroleum Computer Conference held in Dallas, Texas, U.S.A., 31 July-3 August 1994.
48. Yeboah, Y.D., Somuah, S.K. and Saeed, M.R.: "A New and Reliable Model for Predicting Oilfield Scale

Formation," SPE paper No. 25166 presented at the SPE International Symposium on Oilfield Chemistry, New Orleans, Louisiana, 1993.

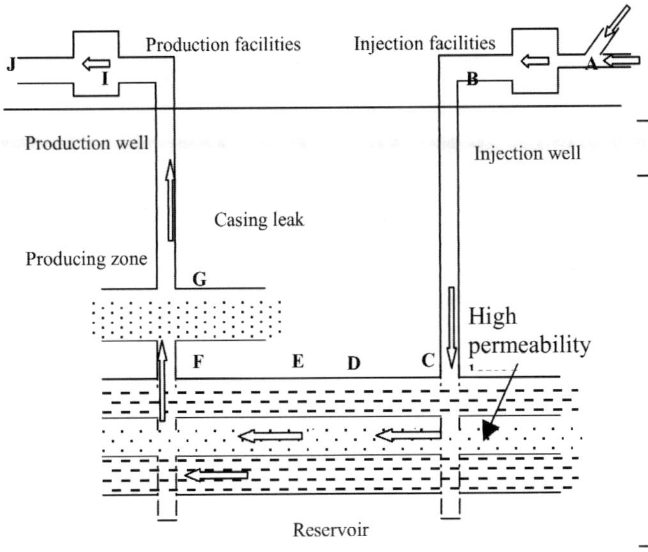

Location	Change which could produce scale formation
A to B	Mixing of brines for injection
B to C	Pressure and temperature increase
C to D	Pressure decline and continued temperature increase
C to F	Solution composition may be adjusted by cation exchange, mineral dissolution or other reactions with the rock
D to F	Mixing of brines in the reservoir
E to J	Pressure and temperature decline. Release of carbon dioxide and evaporation of water due to the pressure decline if a gas phase is present or forms between these locations.
F	Mixing of formation water and injection water which has "broken through" at the base of the production well
G	Mixing of brines produced from different zones.
H	Mixing of produced brine with brine from casing leak

Fig. 2 Diagram indicating changes which could produce scale at different locations

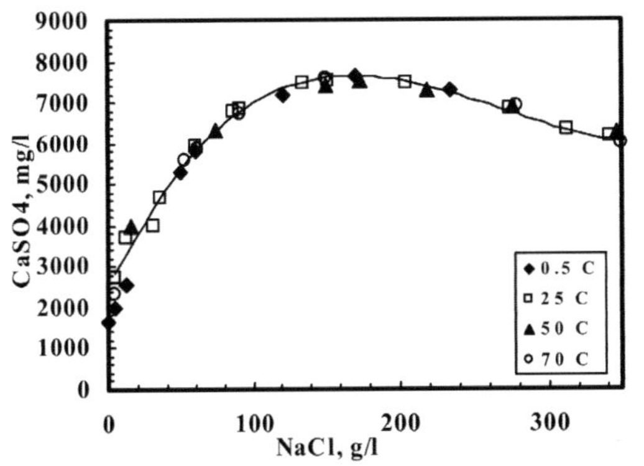

Fig. 3 Solubility of gypsum in NaCl brines at temperatures 0 to 70 °C

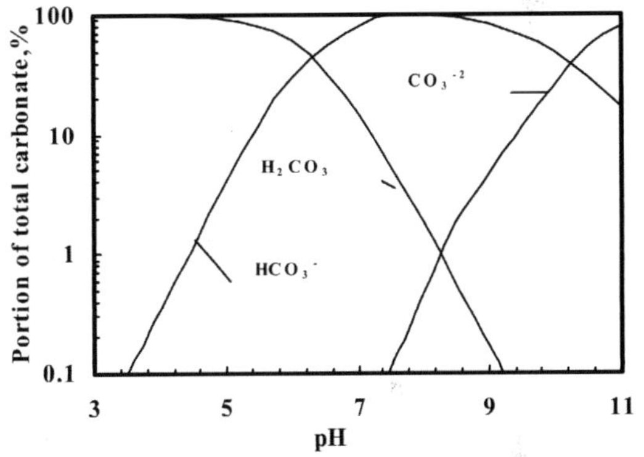

Fig. 4 Ionisation of carbonic acid at different pH values

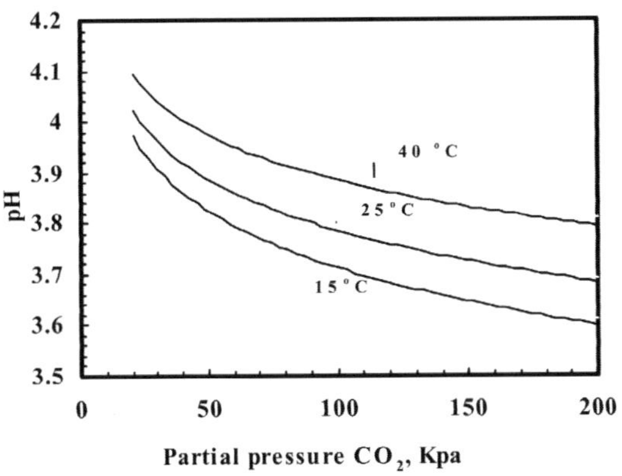

Fig. 5 Effect of CO_2 partial pressure on pH of water

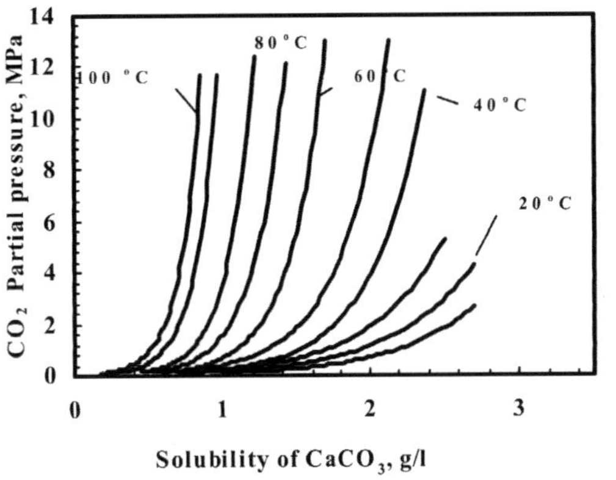

Fig. 6 Effect of CO_2 partial pressure on $CaCO_3$ solubility

1 Holding tanks
2 Stirrer
3 Band heaters
4 Rotameter
5 Water cooler
6 Peristaltic pump
7 Thermocouples
8 Pressure transducers
9 Stainless steel test section
10 Perspex test section
11 Burette
12 Flowmeter
13 Data acquisition
14 Personal computer
15 Printer
--- Electrical line
⋈ Valve
∇ Drain
► Check valve
— Liquid line

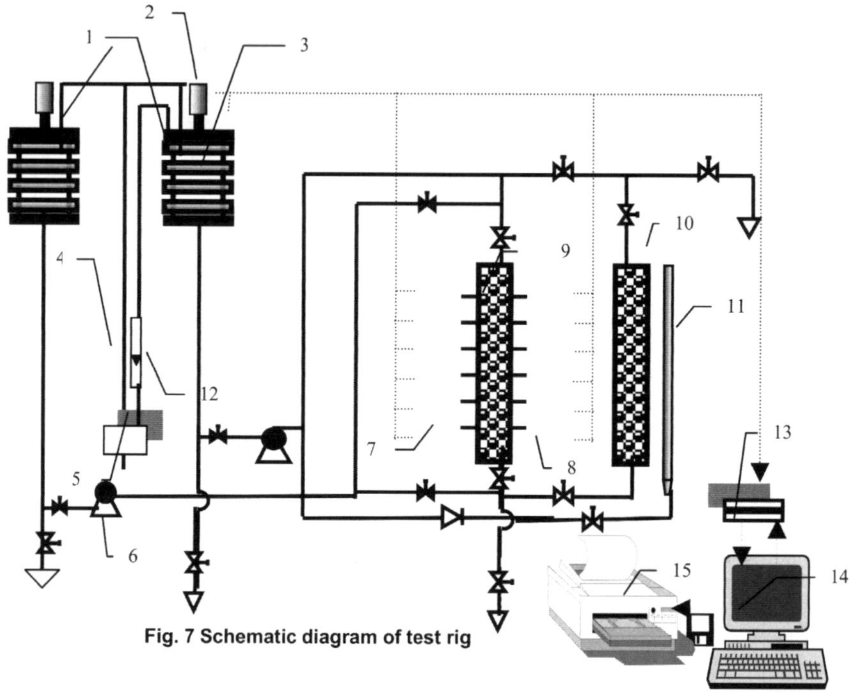

Fig. 7 Schematic diagram of test rig

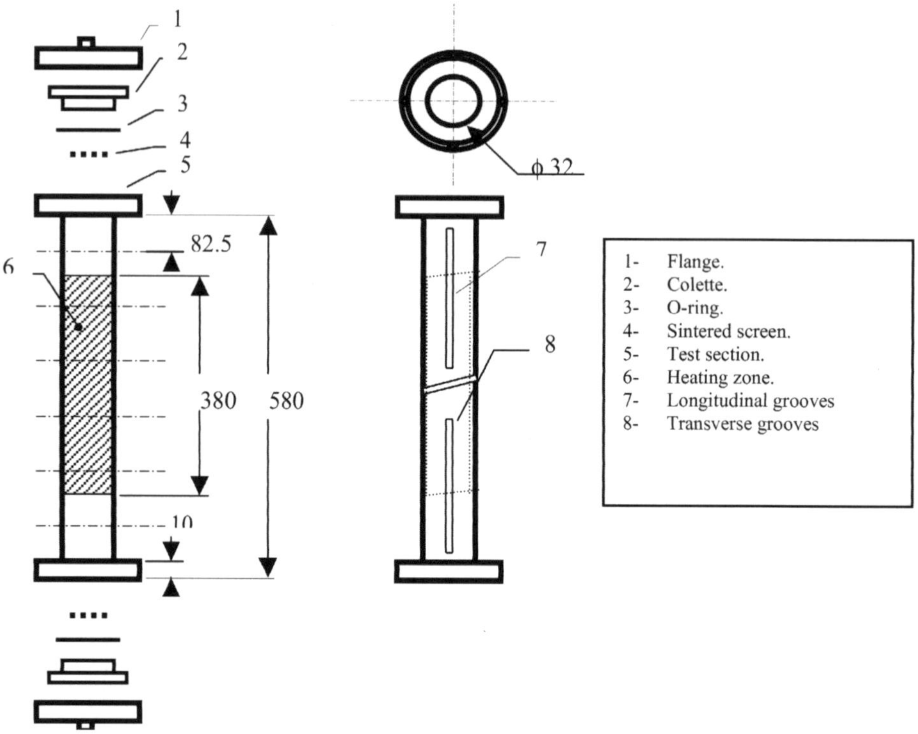

Fig. 8 Design details of the test section.

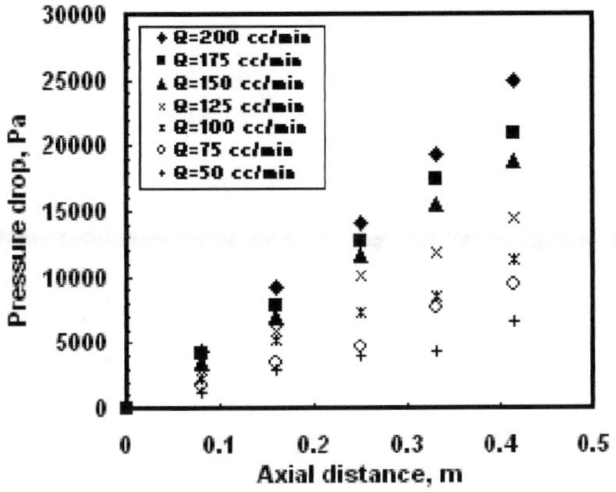

Fig. 9 Variation of pressure drop as a function of distance

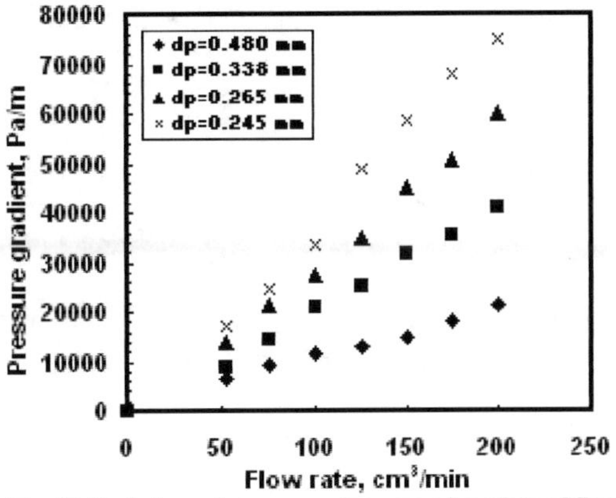

Fig. 10 Variation of pressure drop as a function of flow rate

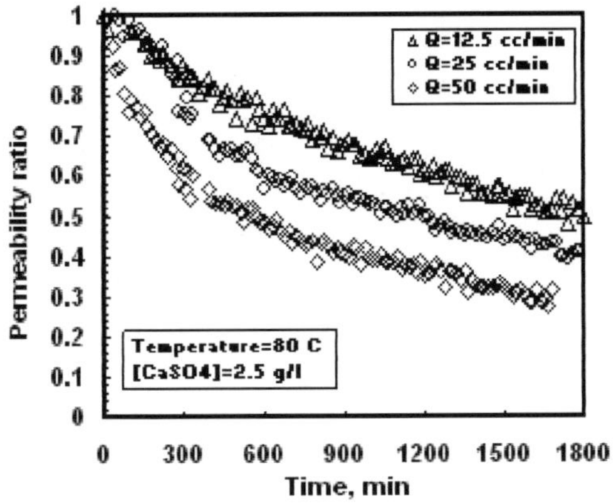

Fig. 11 Variation of permeability ratio as a function of time

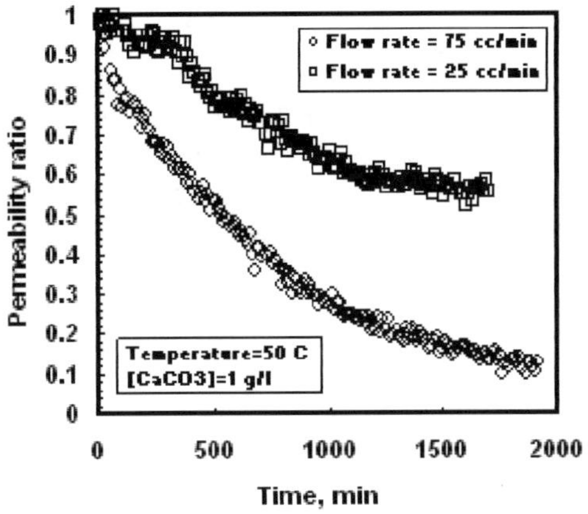

Fig. 12 Variation of permeability ratio as a function of time

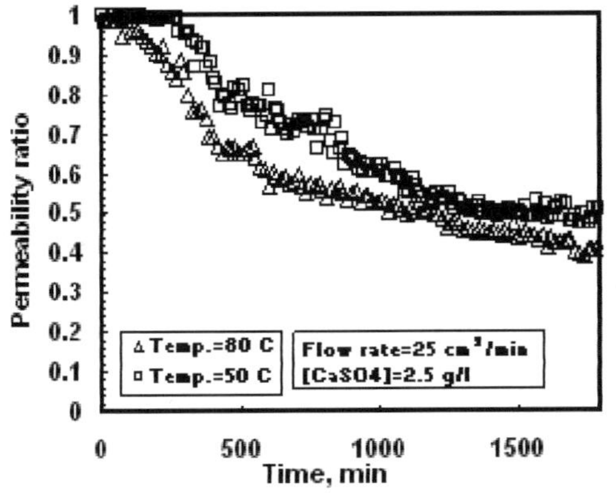

Fig. 13 Variation of permeability ratio as a function of time

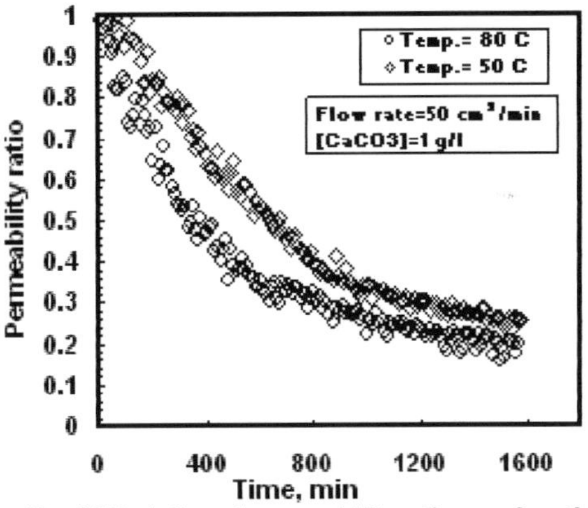

Fig. 14 Variation of permeability ratio as a function of time

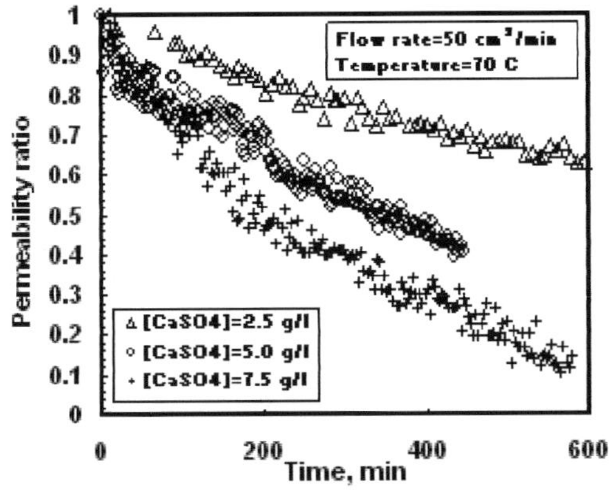

Fig. 15 Variation of permeability ratio as a function of time

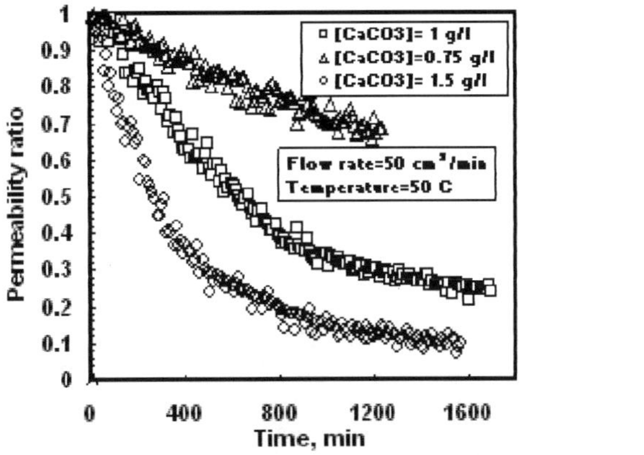

Fig. 16 Variation of permeability ratio as a function of time

Hydrogen and fuel cell – Key to the future clean energy systems

Ken-ichiro Ota

Department of Energy and Safety Engineering, Yokohama National University
79-5 Tokiwadai, Hodogaya-ku, Yokohama-shi, Kanagawa 240-8501, Japan

2004 Prepared for Presentation at 2004 Spring National Meeting/ April 26, 2004/ Tutorial – Hydrogen and Fuel Cell – Key to the Future Clean Energy Systems

Abstract The global carbon cycle is changing by the emission of CO_2 with large consumption of fossil fuels. The global environment including human society should stand on harmonizing with the Earth where the global recycles of materials are important. The hydrogen-based energy system using renewable energy is discussed as an ideal energy system based on global water cycle, which could steadily dispose the entropy to out of Earth. The fuel cell technology is a key for the hydrogen-based energy system. The characteristics of fuel cells are reviewed, especially on the polymer electrolyte fuel cells for vehicles and stationary use.

1. Introduction

Now when we stand at the beginning of the 21st century, human beings have several serious problems to maintain the activity of the civilization. In particular to overcome an energy crisis and to construct a clean energy system have to come true immediately. Although these issues are generally called energy problems, they should be just called "entropy problems". The viewpoint of entropy is very important when energy and environment are discussed, because all activities on Earth including the human beings cause the entropy production according to the second law of thermodynamics with the conservation law of materials and energy. Entropy governs a direction of all changes, so that the entropy production should be considered in order to design the sustainable human society. There are several studies in consideration of entropy when an ideal energy system of civilization is planned [1,2]. However, little has been mentioned the relation of entropy disposal and constancy of the civilization. Therefore, we first discussed the importance of material circulation from the viewpoint of entropy. The hydrogen-based energy system using renewable energy is discussed as an ideal energy system based on water cycle of Earth, which could steadily dispose the entropy produced in civilization to out of Earth. The fuel cell technology is a key for the hydrogen energy system. The characteristics of fuel cells are reviewed, especially on the polymer electrolyte fuel cells for vehicles and stationary use.

2. Relation of Earth and human society considered from the entropy disposal

It was pointed out that the viewpoint of entropy was important when resources and environments were discussed [3]. The purpose of the human society is to maintain the stability itself. According to the second law of thermodynamics, any activity on the Earth always increases total entropy. The value of the entropy which human society produced must be the constant in order to maintain its steady state. It is necessary to import materials and energy of low entropy (resources) from the outside so that the mankind may carry out the activity inside its society. Then, using incoming resources, various activities are performed in the human society, and the entropy is produced with it. Because total entropy never decreases, the entropy, which was produced in civilization must be exported to the outside (environment).

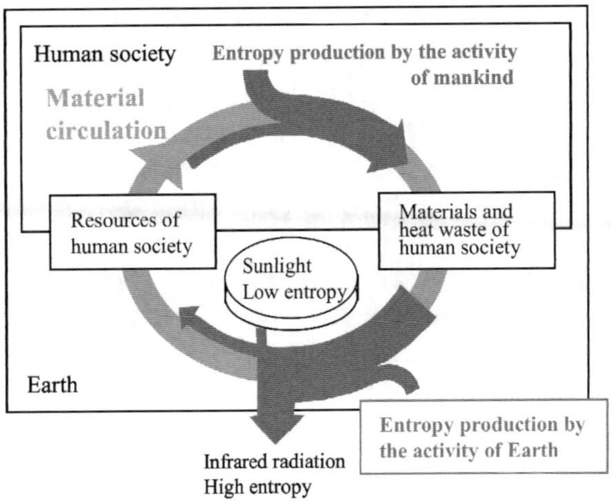

Fig.1 Entropy flow of human society and earth for sustainable society.

It is necessary that the environment also exports the entropy released from the human society outside to maintain its steady state. The steady state of the civilization in the environment cannot be maintained, when the produced entropy is accumulated in the environment. Regarding energy, the Earth is always supplied with the sunlight of low entropy, and the infrared radiation of high entropy can be disposed of to space. Therefore, regarding energy it is possible to maintain the steady state by the flow from the sunlight to the long-wave thermal radiation. On the other hand, regarding materials, the Earth has closed almost. Therefore, the environment must have the mechanism, which returns a material waste of the human society to material resources of the human society in order to constantly supply of the material resources of low entropy from the environment and dispose of a waste of high entropy to the environment. The material entropy can be decreased by converting into a heat. By utilizing the sunlight of low entropy, it is possible that the high entropy with material wastes can be wiped out as infrared radiation to space. As a result, materials circulate. This situation was schematically shown in Fig. 1. Materials permanently circulate, and produced entropy is always disposed of as a long-wave thermal radiation to space. This is an ideal steady state of Earth including civilization. Therefore, it is desirable that the ideal energy system of human society is also based on materials circulation to which include Earth.

3. Ideal energy system consisted on material circulation

The primary energy, initially fossil based, progressively shifts toward renewable sources. It is assumed that the renewable energy is primary energy of ideal energy system. As a material circulation which mankind can utilize as energy system, water cycle is compared with carbon cycle. The water cycle plays an important role for the entropy disposal of Earth at present. Abundance and circulation of the water on Earth [4,5] are shown in Fig.2. An amount of evaporation of present water is for 496T ton/year, and it is equal to amount of rainfall, and it circulates in respect of the atmosphere in about 10 days. On the other hand, carbon cycle using biomass and so on

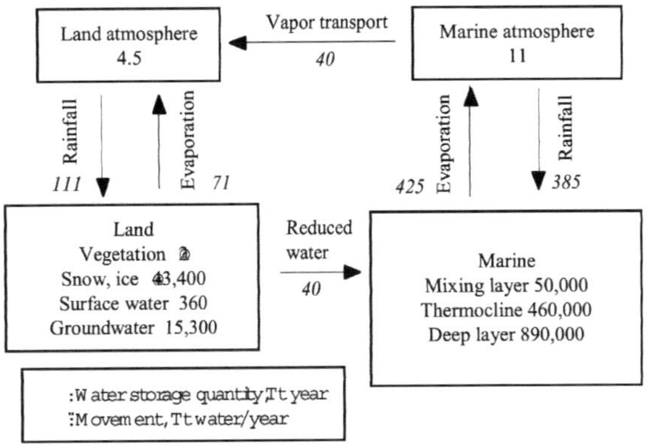

Fig.2 Schematic diagram of water cycle on Earth.

is considered. Figure 3 has shown carbon cycle [6] in present Earth. Present carbonic atmosphere and movement between surfaces are for the 157G ton/year, and it circulates in respect of the atmosphere for about 5 years. It can be said that it discharges the carbon in the atmosphere, since the mankind largely consumes the fossil fuel at present, and the CO_2 concentration increases. The mankind should choose the circulation possessed with Earth of which the effect of mankind on original circulation is small.

Table 1 shows the comparison of carbon and water on Earth. The total abundance of water is much greater than that of carbon. 26,000 times are different in the weight ratio. The water abundance in the atmosphere is also more abounding than carbon abundance in the weight ratio about 21 times. Circulation with much abundance in the nature is difficult to be affected by human activity. The annual movement from the atmosphere of water is 3160 times larger than that of carbon, so that a large difference occurs in the mean residence time in the atmosphere. In comparison with carbon cycle, the water cycle is difficult to receive the effect of the mankind activity. The ratio of energy consumption of mankind to annual movement from atmosphere of water is more than two orders of magnitude less than that of carbon. Water cycle is superior to carbon cycle as material circulation which the mankind utilizes for energy system. The hydrogen-based energy system is the water cycle society. From the viewpoint of entropy, hydrogen-based energy society is the ideal energy system, which the human should pursue. Figure 4 shows ideal hydrogen-based energy system with water cycle.

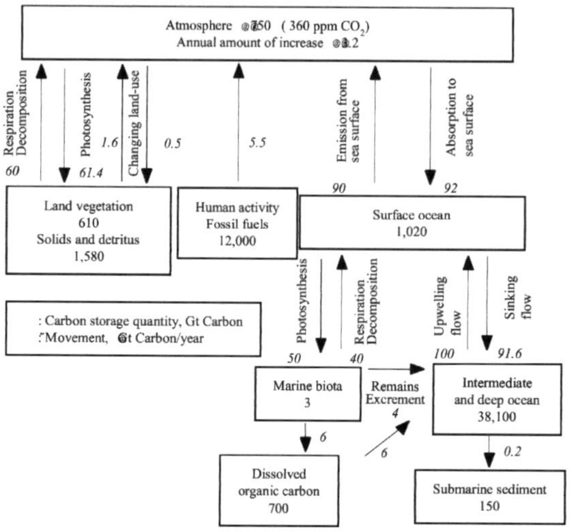

Fig.3 Schematic diagram of carbon cycle on Earth.

Table 1 Comparison of cycle quantity of carbon and water.

	Carbon	Water
Total amount	54,000Gt Carbon	1,400,000Tt Water
Atmosphere abundance	750Gt Carbon	15,500Gt Water
Annual movement from atmosphere	157Gt Carbon	496,000Gt Water
Average retention period in atmosphere	5 year	10 day
Ratio of energy consumption of mankind to atmosphere abundance	0.7%	0.4%
Ratio of energy consumption of mankind to annual movement from atmosphere	3.5%	0.01%

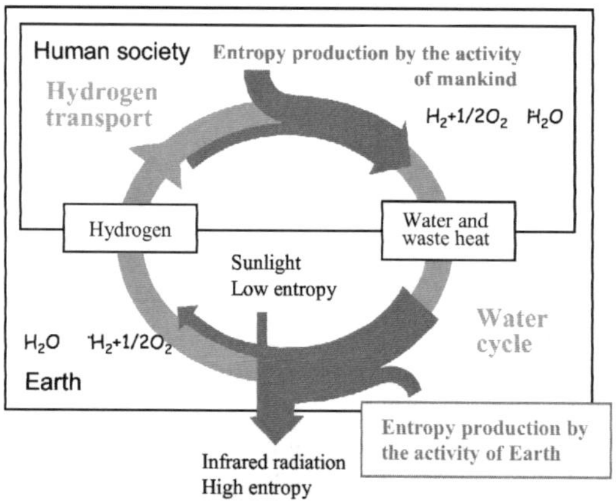

Fig.4 Hydrogen-based energy system with water cycle.

4. Role of fuel cell in hydrogen-based energy society

Fuel cells are theoretically very efficient energy conversion system from hydrogen or fossil fuels to electrical energy. The fuel cell was first demonstrated by Sir W. Grove in1839 [7], using sulfuric acid solution and Pt catalyst. In Japan, Tamaru reported the electricity production at the oxidation reaction of carbon in molten carbonate in 1935 [8,9].

The most popular reaction for a fuel

cell is the oxidation of hydrogen by oxygen.

$$H_2 + 1/2\ O_2 \rightarrow H_2O$$

The reaction is exothermic and Gibbs energy decreases with increasing temperature. Figure 5 shows the dependence of the theoretical efficiency of the hydrogen-oxygen fuel cell on temperature. The Carnot efficiency is also plotted. The efficiency of the fuel cell decreases as the temperature increases. On the other hand the Carnot efficiency increases at higher temperatures. The fuel cell should be operated at lower temperatures to obtain the high efficiency.

The characteristics of fuel cells are summarized as follows.

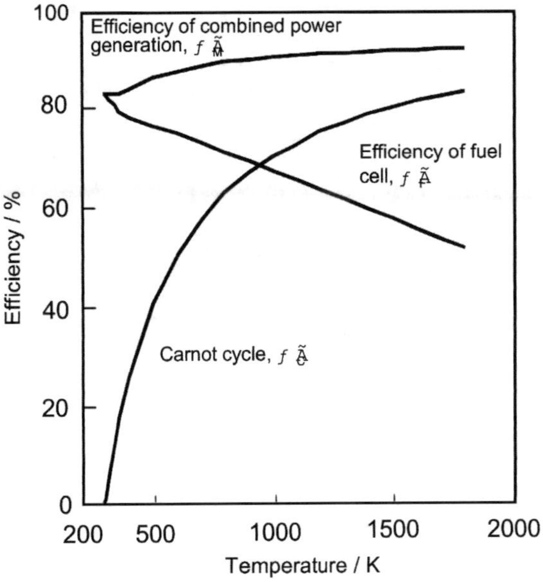

Fig.5 Theoretical efficiency of fuel cell and Carnot efficiency.

(1) The theoretical efficiency of the energy conversion is very high, especially at low temperature.
(2) The scale merit is small. On the other hand, the high efficiency can be obtained even in a small unit.
(3) The exhaust heat can be used easily, that improves the total energy efficiency.
(4) No pollutant (especially NO_x), no noise and no vibration. The environmentally friendly generation system is possible.
(5) The fuel cell is an electrochemical system, which means the reaction at the electrode/electrolyte interface is most important. The electrochemical reactor is a 2 dimensional reactor and the volume efficiency is generally not good.
(6) The cell voltage of a single cell is normally less than 1 V. In order to get a high power density, the reactants (fuel and oxidant) should react rapidly at the electrode/electrolyte interface.
(7) Total system is very expensive, compared to a conventional electric generation system or an internal combustion engine.

5. Fuel Cells for distributed power system

Fuel cells can be operated from room temperature to 1000°C, depending on the electrolytes. Table 2 shows the types of fuel cells and their main characteristics. Figure 6 shows the typical I-V characteristics of the fuel cells. Fuel cells can be used for power generation for vehicles and mobile goods. In particular, polymer electrolyte fuel cells (PEFCs) are expected to be applied to the transportable applications, especially to the automobile use, due to its high power density and low operating temperature. A polymer electrolyte fuel cell was originally developed for the space use. Unfortunately, alkaline fuel cells have been used for the space programs. In the middle of 1980, Dow Chemical developed a new polymer membrane that shows the higher ionic

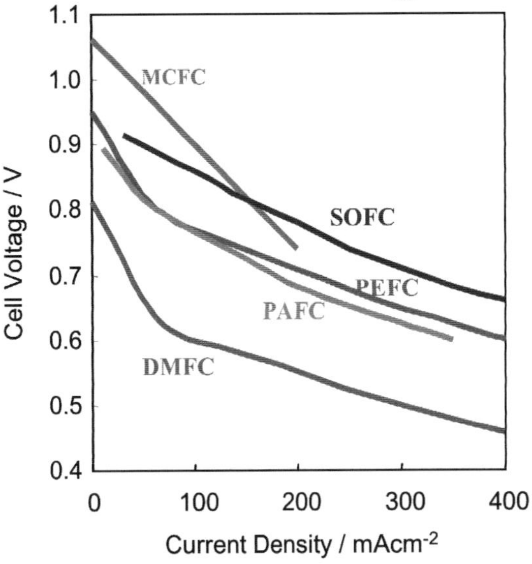

Fig.6 I-V Characteristics of Fuel Cells. (1 atm, U_f=70 85%)

Table 2 Type of fuel cell.

	Alkaline (AFC)	Phosphoric acid (PAFC)	Polymer electrolyte (PEFC)	Molten carbonate (MCFC)	Solid oxide (SOFC)
Electrolyte Carrier ion	KOH aq OH^-	H_3PO_4 aq H^+	Polymer electrolyte H^+	Li_2CO_3/Na_2CO_3 CO_3^{2-}	Stabilized ZrO_2 O^{2-}
Operating temp.	5 240°C	160 210°C	60 80°C	600 700°C	900 1000°C
Fuel	H_2	H_2	H_2	H_2, CO	H_2, CO
Main features	Low operating temperature Wide selectivity of materials	Compact system Effective use of waste heat	High power density Low operating temperature	High cell voltage High total efficiency	Internal reforming is possible.

conductivity than Nafion®. This allows the operation to be a high power density. Most of the automobile makers are interested in this system and developing very hardly. PEFCs have following merits.

1. Since PEFCs are operated at low temperature, it is easy to start or stop and the theoretical efficiency is high.
2. Polymer membrane is very flexible and a PEFC can be constructed in either form.
3. High power density is possible using a thin polymer membrane.

On the other hand, PEFCs have following demerits.

1. Pt electrocatalyst is inevitable at low temperature operation and is very sensitive to the CO content in the anode gas.
2. Water management is very difficult to handle.
3. High cost
4. Although the theoretical efficiency is high, the actual efficiency is not so high. We have to improve the efficiency.

PEFCs are developing for a home cogeneration system as well as the automobile application. For a home cogeneration a smaller size (- 1 kW) is considered in Japan. In such a small scale, fuel cells have a highest efficiency among the power generation technology.

Considering the characteristics of electrochemical systems, fuel cells can be applied to smaller applications. Since the electricity and the thermal energy are used simultaneously, the efficiency for the electricity production will be able to get 30 %(HHV) and the total energy efficiency of 60 %(HHV) or more might be possible. No other system can produce electric energy so efficiently in such a small unit.

6. Conclusion

The global environment including human society should stand on harmonizing with the Earth where the global recycles of materials are important. The hydrogen-based energy system using renewable energy is discussed as an ideal energy system based on global water cycle, which could steadily dispose the entropy to out of Earth. Fuel cell technology is a key for the hydrogen energy system. Fuel cells have theoretically high efficiency especially at low temperatures. This high efficiency does not decrease in a small units. These characteristics are suitable for a distributed power source in small scale. In order to get a real commercialization, there are several problems to be solved, such as cost, durability, efficiency etc. Further developments might be necessarily especially for new electrode materials and new electrolyte materials. Within 21st century, I hope that fuel cell age will come true in many fields from small size to large size.

References

[1] M. A. Rosen and D. S. Scott, *Int. J. Hydrogen Energy*, **28**, 1307 (2003).
[2] M. A. Rosen and D. S. Scott, *Int. J. Hydrogen Energy*, **28**, 1315 (2003).
[3] A. Tsuchida, Shigen Butsurigaku Nyumon (Introduction to Resources Physics), Nippon Housou Kyokai (1982) in Japanese.
[4] Baumgartner and E.Reichel, The World Water Balance, Elsevier Scientific Pub.co. (1975).
[5] E.K.Berner and R.A.Berner, Globlal water cycle, Geochemistry and Environment, Prentice-Hall, Inc., Englewood Cliffs, New Jersey (1987).
[6] D.S.Schimel, *Global Change Biology1*, 77-91(1995).
[7] W.R. Grove, *Phil.Mag.*, **21**, 417 (1842).
[8] S.Tamaru, K.Ochiai, *Nippon Kagakukaishi*, **56**, 92 (1935).
[9] S.Tamaru, K.Ochiai, *Nippon Kagakukaishi*, **56**, 103 (1935).

Wind Energy – A Natural and Significant Source of Electricity

R. Gerald Nix and Brian Smith
National Renewable Energy Laboratory
1617 Cole Blvd
Golden, Colorado 80401

What is Wind Energy?

Wind energy is the capture of the kinetic energy in atmospheric air movements and conversion to torque used to drive a generator to produce electricity or to perform work. At it's simplest; the wind turns the turbine's blades, which spin a shaft connected to a generator that makes electricity. Large turbines can be grouped together to form a wind power plant, which feeds power to the electrical transmission system.

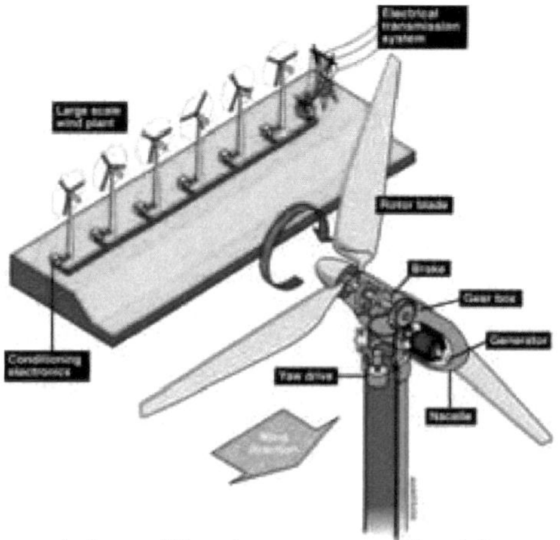

Wind energy is an intermittent resource that is very widespread. Each geographic area has a long-term average characteristic wind pattern, with the primary discriminator being the annual average wind speed that defines an energy class for the winds of that area. There is a significant variation in the instantaneous value of the energy in the wind, significant diurnal variation and inter-annual variation. A wind resource map of the United States is shown below:

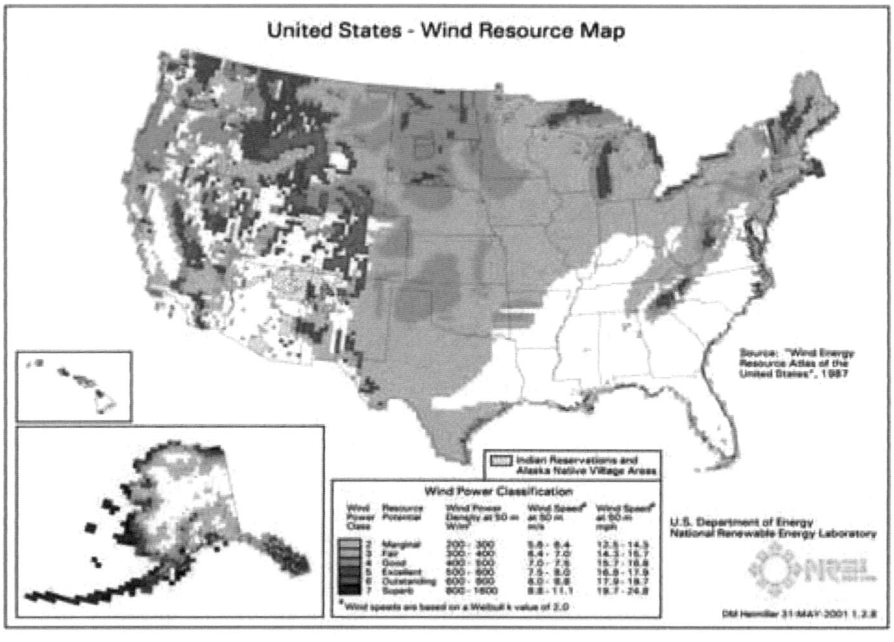

Wind Energy is Growing and Is a Significant Source of Electricity

Wind energy has the potential to significantly contribute to the electricity

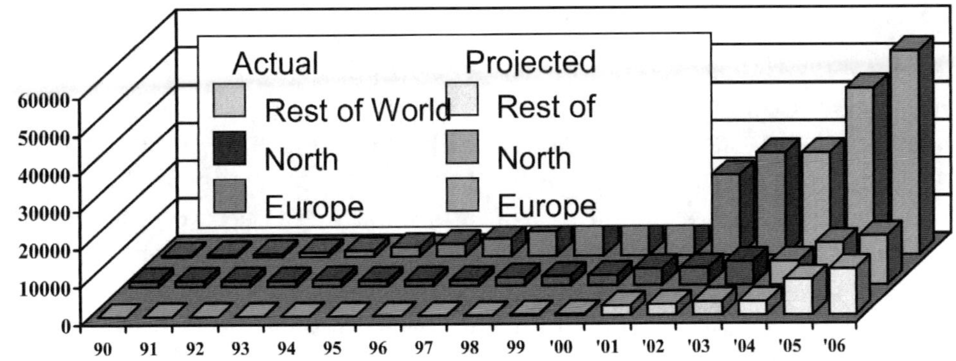

generation capacity within the U.S., with a reasonable goal being to supply 10 percent or greater from wind. The graph below shows the growth of wind energy within various areas of the world for the years of 2000 to 2010 (projected). Wind energy is the fastest growing method of electricity generation within the world.

What is the Status and Challenges Wind Energy?

Wind energy has been under intense development from the 1970s to present, although it has been practiced for uses other than electricity generation for many hundreds of years. It is a maturing technology that comes with its own set of specific challenges for further increasing efficiency, enhancing robustness and reducing cost to provide a basis for even more widespread application.

The commercial direction is toward much larger wind turbines, as shown below.

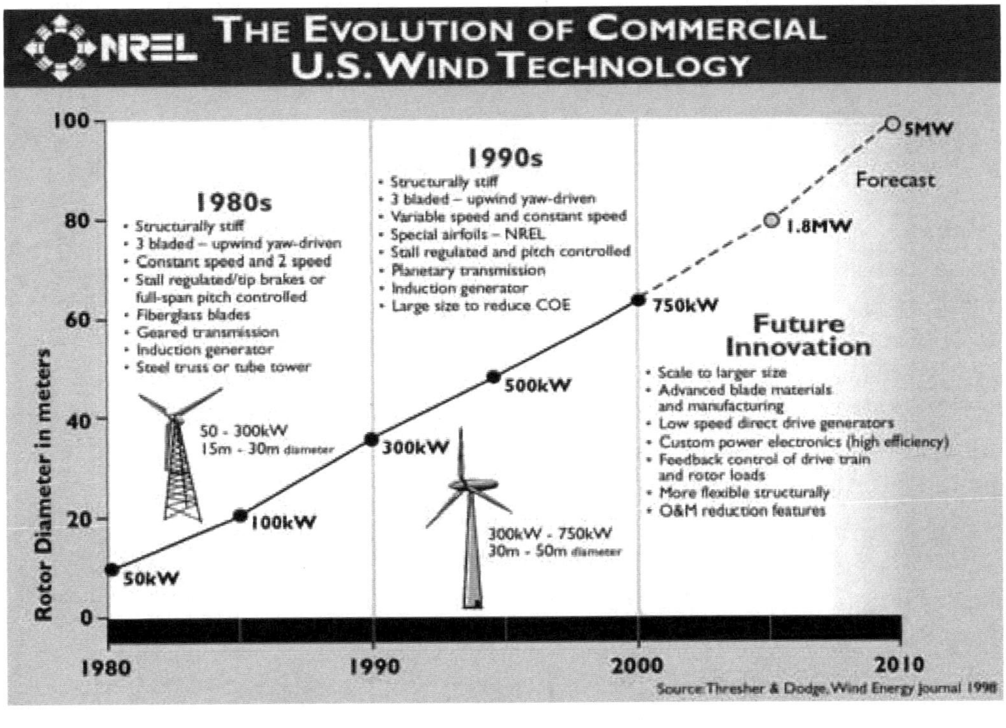

Some of the challenges that will be discussed in detail are those that must be overcome to reach the goals, including:

- Taller Towers – Understanding the Low Level Jet

Wind Energy is proportional to the wind velocity to the cube power, and from basic boundary layer theory, the velocity in the atmospheric boundary layer is proportional to the one-seventh power of the height above the earth's surface. This gives incentive for very tall towers to increase the energy capture – with towers getting sufficiently tall that they are subject to more frequent interaction with a nocturnal jet dropping from higher altitudes as coherent eddies, causing potential significant damage to turbines. This phenomenon occurs when the nighttime cooling allows winds to stabilize into parallel layers flowing at different velocities. At the boundary between these parallel layers, Kelvin-Helmholtz instabilities can form. These instabilities can rapidly grow into large-scale coherent vortices that can reach the ground and can cause significant gust loads on turbine rotors. This interaction must be understood.

The structural aspects of taller towers must be understood to minimize costs of fabrication and erection. Some of the current research and development topics include self-erecting towers and telescoping towers.

- Adaptive Rotors

Wind turbines are usually stall regulated and are subject to a significant variation in lifting forces. To avoid unacceptably large loads, the rotor blades can be pitched or rotated to change lift and loading. Much better is to use designs and materials that passively change shape to become self-limiting with respect to loads. This avoids the expense and operating/maintenance requirements of blade pitch systems. This is a fruitful area for materials and design R&D.

- Integration of an Intermittent Resource

Wind is an intermittent energy source, giving rise to the challenge of how to best make use of the time variant wind-generated electricity within a conventional grid that responds to totally different driving parameters. One extreme is to maintain sufficient spinning reserve within the conventional generating system that the grid is sufficiently stiff that the loss of the entire wind farm output can be accommodated without losing the grid. Research based on statistical analysis of wind data and on forecasting has shown that this is not necessary and that with reasonable geographical distribution, wind farm output is predictable and can be accommodated by the grid without maintaining excessive spinning reserve. This mean that wind energy becomes a somewhat dispatchable generation source as well as a fuel saver when added to a grid based on conventional generation.

- Drive Trains and Generators

R&D is required on innovative generator design to reduce size and cost. This includes the use of better materials, low speed designs and direct drive generators. Concepts include distributed drive trains where the drive path from the rotor is split to drive several parallel generators, each of smaller size. Direct drive generators involve the production of wild alternating current electricity, necessitating the use of power electronics to ensure a steady 60-Hertz output. Design of efficient power conversion and conditioning equipment is a challenge.

- Controls

Wind power is the perfect application for adaptive control, a control system that can adapt to perform optimally despite rotor variability, blade erosion and instantaneous variations in conditions. This is an opportunity for fuzzy logic controllers and for neural network controllers.

- Materials

There are tremendous opportunities for better materials for wind energy applications, ranging from better materials for rotors (example is carbon fiber performance at much lower costs) to towers (perhaps lightweight high-strength concrete) to fabrication techniques (field fabrication of blades to avoid transportation limitations).

- Deployment

One of the obvious opportunities is for offshore collection of wind energy, with transmission back to the grid via underwater cables. Here designers face challenges of the type of platform and the type of anchoring. This gets into innovative designs that enhance robustness of the entire wind energy system and minimizes maintenance requirements since the offshore systems will be more difficult to access than land-based systems.

- Virtual Prototyping

As with any mechanical system, the challenge is to sufficiently understand the physical performance sufficiently that the system can be modeled, the models validated with field data, and the models applied to develop several generations of improvements within the computer, prior to "cutting metal" to build a prototype for testing. This has the potential to significantly reduce development costs.

What's in Wind Energy for Chemical Engineers?

Chemical engineers may consider wind energy as something for mechanical engineers, electrical engineers, structural engineers or aerodynamicists. However, there are significant aspects that do relate to the training and capabilities of chemical engineers, including materials science, manufacturing engineering, controls engineering, atmospheric science and fluid flow, environmental engineering, systems engineering, and others such as business development. There are many potential opportunities and challenges within wind energy that can be solved using the tools in the chemical engineers toolkit, particularly when working as a member of an interdisciplinary team.

Biological Processing of Cellulosic Biomass to Fuels and Chemicals: Progress, Challenges, and Research Opportunities

Charles E. Wyman
Thayer School of Engineering, Dartmouth College
8000 Cummings Hall, Hanover, NH 03755

ABSTRACT

Cellulosic biomass including agricultural and forestry residues, portions of municipal solid waste, and herbaceous and woody crops provides a unique resource for sustainable production of organic fuels and chemicals, and the cost of many cellulosic materials is competitive with petroleum. Biological processes offer the potential for economical conversion of cellulosic biomass to such products and can result in very low greenhouse gas emissions. Substantial progress has been made toward competitiveness. However, capital costs are high for initial projects, and the perceived risk of commercializing new technology presents a major obstacle to commercial use. Thus, it is important to take advantage of niche applications that can reduce costs. In addition, we must improve the technology platform to support scale up with confidence. For example, although pretreatment by hemicellulose hydrolysis has been modeled based on only reaction considerations, mass transfer appears to play an important role, and a better understanding of such phenomenon can accelerate commercial use and suggest advanced technologies to improve costs.

KEY WORDS

biomass, ethanol, chemicals, energy, environment

INTRODUCTION

Through continued funding from mostly the federal government, technology for breaking down cellulose and hemicellulose in woods, grasses, and other inexpensive forms of cellulosic biomass to sugars has progressed significantly from early acid systems such as the Scholler process of the 1930s and the Madison process of the 1940s to the point that companies seek to commercialize enzyme based processes for production of fuels and chemicals. Enzymes offer the near theoretical yields needed for long term economic viability, and modern biotechnology that emerged in the mid 1970s provides a powerful tool for enhancing performance and reducing costs. As one example, bacteria were genetically engineered within the last 15 years to ferment all 5 sugars in biomass to ethanol with high yields, significantly enhancing revenues from the substantial quantities of arabinose and xylose in many forms of biomass that would be previously wasted and entail large disposal costs. Thus, although some may judge research expenditures to be high, progress has been significant and taken us to the edge of commercial use. The needs are to commercialize the technology and further improve it.

THE CHALLENGE

Unfortunately, commercialization of new technology is daunting, and its perils are not well understood. For example, investors demand very high rates of return to compensate for the risk of introducing technology for the first time. Furthermore, commodity products such as ethanol must compete in markets with very thin margins. Additionally, economies of scale dictate large plants be built, increasing costs and concerns about risk and adequate returns. Competing products, e.g., gasoline for ethanol, are well entrenched in terms of manufacturing technology; pay off of capital investments; marketing, distribution and use infrastructure; policy; and acceptance. The costs of producing gasoline and many other petrochemical products also benefit from over 100 years of learning curve experience with huge cumulative production. These and many other factors represent major hurdles to introducing new commodity products.

THE NEED

Although introducing fuels and chemicals production from biomass is very challenging, the payback is large. First, the largest fraction of energy we use, about 40%, is from petroleum, and although domestic production and consumption of all other energy sources are roughly balanced, we import about 60% of our oil. Furthermore, most petroleum is located in less stable regions that are in continual upheaval, making cost and supply disruptions a historic reality and an ongoing danger. With about 2/3 of petroleum used for transportation that is over 96% dependent on oil, it is obvious that we must do something about transportation energy if we hope to affect our oil dependence. Beyond such security concerns, the United States releases ~25% of world greenhouse gases each year, with about one third from transportation. Thus, it is highly desirable to introduce a transportation fuel that is domestically secure and emits little if any greenhouse gas emissions. Making ethanol from cellulosic biomass is a powerful and unique route to meeting these needs from an inexpensive and sustainable resource. Furthermore, cellulosic biomass could provide a valuable substrate for making organic chemicals that are so important to our standard of living.

BIOLOGICAL PROCESSING OF CELLULOSIC BIOMASS

The general biological processing scheme starts with milling and pretreatment of cellulosic biomass to overcome its natural resistance to biological degradation, with dilute acid hydrolysis of hemicellulose often used to achieve hemicellulose sugar yields of up to 90% and make highly digestible cellulose. Liquid hydrolyzate is treated to remove or inactivate compounds naturally released (e.g., acetic acid, lignin fragments) or formed by degradation (e.g., furfural) that are inhibitory. A small portion of the pretreated solids and possibly some liquid hydrolyzate is used for cellulase production by the fungus *Trichoderma reesei* or other organism. This cellulase is added to pretreated solids to breakdown cellulose to glucose with yields of ~90% possible. The glucose can be fermented or chemically reacted to various products, with yeast fermenting it to ethanol at ~90-92% of theoretical yields, as an example. Next, product is recovered, with distillation columns and molecular sieves often used in the case of ethanol. The lignin and other unconverted materials can be fed to a boiler to provide heat and electricity for the entire process with excess electricity sold, or the lignin can be converted to other products. The liquid not retained with the solids is treated, with clean water discharged or recycled to the process and sludge and methane are fed to the boiler. Ash is landfilled.

PRETREATMENT OF CELLULOSIC BIOMASS

Pretreatment improvements to enhance sugar yields from cellulose and hemicellulose have been important to cost reductions. In brief, various biological, chemical, and physical pretreatments have been applied over the years, and ammonia, lime, controlled pH, sulfur dioxide, and dilute acid pretreatments are receiving continued attention. Dilute sulfuric acid pretreatment has been the subject of considerable research and development, particularly targeted at fuels production. While sugar yields from uncatalyzed steam explosion are limited, Grethlein and Converse at Dartmouth showed that dilute sulfuric acid dramatically improved performance at a relatively low cost, and the process has been demonstrated to be effective on a variety of biomass feedstocks. For example, 85-90% or more of the sugars can be recovered from the hemicellulose fraction, and about 85-90% of the remaining cellulose can be enzymatically digested to glucose. These high yields are vital to low costs and have led many such as the National Renewable Energy Laboratory (NREL), overseers of the largest biomass ethanol development effort in the world, to favor dilute sulfuric acid hydrolysis of hemicellulose for near term applications.

CELLULASE PRODUCTION

Various publications provide more information on cellulase production than can be covered here. In summary, the fungus *Trichoderma reesei* or strains derived from it are typically used for cellulase production. Some cellulase is now produced commercially for low volume, high value specialty markets such as stone-washed jeans that command high prices. Cellulase production research for fuels and commodity chemicals has been very limited, and analyses have projected costs of ~$3.00/gallon if cellulase is purchased or $0.30 to 0.50/gal of ethanol produced if cellulase is manufactured on site. These costs are higher than estimated in earlier studies, showing the uncertainty in cellulase production technology.

Inducers such as lactose or cellulose promote cellulase release at high yields. Production on mixed solids/liquid hydrolyzate from pretreatment promises to reduce cellulase costs and simplify the integrated system. Adding the entire cellulase broth to the enzymatic hydrolysis step improves performance some because fungal bodies retain enzyme and particularly beta-glucosidase activity. This approach also eliminates equipment, reduces the opportunity for microbial invasion, and passes unused substrate for conversion to product, increasing yields.

About 3 years ago, the U.S. Department of Energy Biomass Program awarded over $30 million to Genencor International and Novozymes to lower cellulase costs by more than a factor of ten. These companies applied their extensive biotechnology tools to increase specific activity, reducing cellulase loadings to achieve a target cellulose conversion. They also developed improved production techniques such as in situ generation of the inducer sophorose that makes it possible to use soluble sugars for enzyme production and eliminate insoluble inducers such as cellulose while maintaining high enzyme yields. The result likely includes higher productivity and lower mixing power requirements with significant cost reductions. The current cost claims of ~$0.35-0.50/gallon of ethanol produced are a significant step but are predicated on much greater cellulase production than currently practiced, poising a substantial chicken-and-egg challenge for first projects. Costs of <$0.10/gallon of ethanol that are more in line with corn ethanol enzyme costs would be much more conducive to commercialization.

CELLULOSE HYDROLYSIS

Acids or enzymes known as cellulases catalyze reaction of water with the glucan molecules in cellulose, and 180 mass units of glucose are released from 162 mass units of glucan and 18 mass units of water, an 11.1% mass gain. Oligomers, often containing only 2 to perhaps 3 glucose units, may also be released as intermediates in cellulose hydrolysis. Because cellulase enzymes are very specific and the optimum temperature is only about 50°C, degradation reactions are virtually eliminated, and glucose yields can approach 100%. On the other hand, dilute acids (e.g., 1.0% sulfuric acid) require much higher temperatures of ~220°C that also trigger glucose degradation to such products as hydroxymethyl furfural, limiting yields. Concentrated acids (e.g., 75% sulfuric acid) can be used at moderate temperatures to achieve high yields similar to those for cellulase enzymes but must be recovered at low costs.

The major challenge in cellulosic biomass conversion is to improve hydrolysis of recalcitrant cellulose, the largest single fraction of biomass. In fact, most of the historic cost reductions resulted from improving dilute acid pretreatment and enzymatic hydrolysis of cellulose with cellulase from strains of the organism *Trichoderma reesei* discovered during World War II. This fungus evolved through classical mutations and selection from earlier strains such as QM9414, to improved varieties such as Rut C30, and on to Genencor 150L that was more effective because of enhanced levels of beta-glucosidase that convert cellobiose, a particularly powerful inhibitor of cellulase, into glucose.

SUGAR FERMENTATION

The sugars released from hemicellulose and cellulose can be converted into a variety of products by biological or chemical means. Native organisms do not ferment the five sugars released through hemicellulose hydrolysis (arabinose, galactose, glucose, mannose, and xylose) to ethanol with high yields. However, in the late 1980s, researchers at the University of Florida were able to insert two genes from the bacteria *Zymomonas mobilis* into *E. coli* and other bacteria that naturally take up all five sugars and redirect the intracellular intermediate pyruvate into ethanol instead of the products that it naturally produces. The resulting organisms realize nearly theoretical yields of ethanol from all biomass sugars, a major step toward commercial success.

In the mid 1970s, Takagi et al showed that combining fermentation to ethanol with hydrolysis via simultaneous saccharification and fermentation (SSF) reduced accumulation of glucose, another powerful cellulase inhibitor. Even though the temperature had to be reduced somewhat from the optimum for cellulose hydrolysis, SSF gave better rates, yields, and concentrations of ethanol than hydrolysis alone. Nonetheless, cellulase action was still slow, with typical SSF reaction times of about 5 to 7 days to achieve modest ethanol concentrations, and fermentative organisms were sought that tolerate the combined stresses of higher temperatures (to increase hydrolysis rates), low glucose levels (due to rapid sugar metabolism by the organism), and high ethanol concentrations. Because rapid conversion of cellobiose to glucose was more important than temperature, supplementation with beta-glucosidase was needed to keep cellulase loadings reasonable. Subsequent research showed that the yeast *Brettanomyces custersii* that directly ferments cellobiose into ethanol enhanced performance, and bacteria that ferment cellobiose to ethanol should provide similar benefits, as should organisms that also make cellulase components. SSF has been applied to a many feedstocks pretreated at various conditions, and temperature tolerance and other traits that better match preferred operating conditions for cellulase are still being sought. Combining SSF with hemicellulose sugar fermentation has received attention to lower costs, and SSF has been used to make products other than ethanol. SSF has also been applied to fed-batch and continuous processes for conversion of paper sludge and wood. Although other approaches are still considered, SSF technology will likely remain in the lead, at least for making ethanol, until cellulases are developed that work much faster with minimal product inhibition at high temperatures.

PRODUCT RECOVERY

Ethanol can be concentrated to near its azeotrope by conventional distillation with molecular sieves used to remove the remaining water. This technology is tried and true and, contrary to many reports, low in capital and operating costs. On the other hand, recovery of low volatility products other than ethanol from a cellulosic fermentation broth that contains solids (e.g., lignin) and minerals (e.g., gypsum) presents important challenges that have not been fully dealt with for many products.

RESEARCH OPPORTUNITIES

Fermentation of all five sugars in biomass to ethanol with high yields has been developed and is no longer an important economic issue. Cost effective technology for recovery of ethanol from the fermentation broth is also in hand. The primary challenge is to reduce the cost of breaking down the hemicellulose and cellulose in biomass to sugars – overcoming the recalcitrance of biomass. In particular, advanced pretreatment approaches are needed that reduce chemical costs, can use low cost materials of construction, do not require hydrolyzate conditioning before fermentation, and insure high yields of sugars from hemicellulose and cellulose.

Cellulase enzymes are expensive, and we must find ways to further reduce their costs. For example, high doses of cellulase are currently needed, and enhancing their specific activity would cut costs. Higher cellulase productivity and yields would also reduce costs. Better pretreatments could reduce costs by improving the accessibility of cellulose to cellulase and through reducing non productive binding of cellulase on lignin and other non reactive components.

Technology advances and commercial applications would be greatly facilitated by improving our knowledge of biomass processing systems. As an example, accurate predictive tools would be valuable in scaling up pretreatment by hemicellulose hydrolysis and in defining opportunities to advance technology. However, current hemicellulose hydrolysis kinetic models are based on first order kinetic assumptions and have changed little since the mid 1940's. Furthermore, these models display inconsistencies that call into question their mechanistic validity such as their failure to account for the observed change in hemicellulose removal with solids concentration and flow through the solids. One would not necessarily expect homogeneous first order reaction kinetics to necessarily describe solid-liquid reactions in which mass transfer, solubility limitations, and non-homogeneous reactions at the solid-liquid interface could play important roles. Our research has shown that mass transfer models adapted from other applications appear to describe hemicellulose hydrolysis as well as conventional reaction-only models and can better explain why reaction rates change with solids concentrations and flow.

SUMMARY

Conversion of cellulosic biomass can provide a powerful route to meeting our need for organic fuels and chemicals on a sustainable basis. Substantial progress has been made in lowering the processing costs to the point that several companies are pursuing commercial applications. However, the challenge is to overcome the perceived risk of introducing new technology and to continue to advance these processes. Improving our knowledge of such systems can be a powerful tool in this quest through providing more confidence in scale up and suggesting key opportunities to reduce costs. Particular focus on understanding and overcoming the recalcitrance of cellulosic biomass can pay the best dividends in producing fermentable sugars at low costs.

ACKNOWLEDGMENTS

We are grateful to each of the following sponsors for supporting our research: the Initiative for Future Agriculture and Food Systems Program of the US Department of Agriculture (award number 00-52104-9663), the National Institute of Standards and Technology (60NANB1D0064), the National Science Foundation Division of Bioengineering and Environmental Systems (BES-9985351), the National Research Initiative Competitive Grants Program of the US Department of Agriculture (2001-35504-10041), the Office of the Biomass Program of the US Department of Energy (DE-FC36-00GO10589 and DE-FC36-01GO11075), and the Thayer School of Engineering at Dartmouth College.

BIBLIOGRAPHY

Energy Information Administration. *Annual Energy Review 2002*. Washington, DC: US Department of Energy; 2003.

Ingram LO, Aldrich HC, Borges ACC, Causey TB, Martinez A, Morales F, Saleh A, Underwood SA, Yomano LP, York SW. Enteric bacterial catalysts for fuel ethanol production. *Biotechnology Progress* 1999, **15**:855-866.

Lynd LR. Overview and evaluation of fuel ethanol from cellulosic biomass: Technology, economics, the environment, and policy. *Annual Reviews Energy Environment* 1996, **21**:403-465.

Lynd LR, Cushman JH, Nichols RJ, Wyman CE. Fuel ethanol from cellulosic biomass. *Science* 1991, **251**:1318-1323.

Lynd LR, Elander RT, Wyman CE. Likely features and costs of mature biomass ethanol technology. *Applied Biochemistry and Biotechnology* 1996, **57-58**:741-761.

Wyman CE, editor. *Handbook on Bioethanol, Production and Utilization*. Taylor & Francis; 1996.

Wyman CE. Biomass ethanol: Technical progress, opportunities, and commercial challenges. *Annual Review of Energy and the Environment* 1999, **24**:189-226.

Wyman CE. Twenty years of trials, tribulations, and research progress in bioethanol technology - Selected key events along the way. *Applied Biochemistry and Biotechnology* 2001, **91-93**:5-21.

Non-thermal plasma for biomass tar removal

S.A.Nair[1], K.Yan[1], A.J.M.Pemen[1], E.J.M.van Heesch[1], K.J.Ptasinski[2], A.A.H. Drinkenburg[2]
1-Faculty of Electrical Engineering, 2- Faculty of Chemical Engineering,
Eindhoven University of Technology, 5600 MB Eindhoven, The Netherlands
Phone: +31 40 247 4494, *Fax:* +31 40 245 0735, *Email:* S.A.Nair@tue.nl

Abstract: Gasification is considered as an efficient way to utilize biomass as an energy source or for fuel production. One of the obstacles for commercial utilization is the production of tars or heavy hydrocarbons along with the producer gas. Tar removal posses a serious challenge. Conventional methods employ catalytic or thermal cracking or scrubbing techniques. The present paper investigates non-thermal plasma as an alternative approach. Experimental results indicate energy requirement to be about 400 J/L for removal of naphthalene from synthetic producer gas at 200 °C. The main pathway for tar removal by pulsed corona at the investigated temperatures is dry reforming by CO_2. For commercial application, the key issue is related to the energy consumption of the process. In addition, another requirement is that fuel gas cleaning should be done at high temperatures in order to integrate with the downstream process equipments. The paper intends to provide an insight into pulsed corona plasma treatment of fuel gas at high temperatures.

Keywords: plasma, biomass gasification, tar, pulsed corona.

1. Introduction

In context of present day energy resources, biomass is considered a promising fuel for the future. Processing of biomass can be done by gasification - for electricity production or for synthetic gas production. Another alternative is by pyrolysis, which yields heavy oil commonly referred to as "bio-oil". Attempts are made to convert this oil as a transportation fuel or for further conversion to synthetic gas. The efficient conversion process is gasification, where the product is a low calorific value gas (if air is used as a gasifying medium). If the focus is for electricity production, then the produced gas is directly fed into a gas engine for electricity generation. However, gasification gas needs to be pretreated before it can be utilized. Undesired components such as heavy hydrocarbons (normally referred as "tars") are present in the gas, which condenses and therefore clogs the downstream equipment. Along with tars, the gas also needs to be cleaned from dust particles. Additionally, for flexible operation of a gasifier, it has to handle various biomasses and not a particular type. In such circumstances, the gas can also be contaminated with sulfur, chlorine etc. which again could cause corrosion problems and deactivate catalysts if present in downstream processing.

The first problem out of the many is to have an efficient tar removal system. Various removal options are investigated. One of them is to focus on addressing the problem at source, i.e. to operate the gasifier at optimum conditions, and to incorporate catalysts within the gasifier. However, such options tend to be fuel specific, which limits the flexibility of operation of gasifier considering the immense variations in biomass fuels. Apart from this, there are other possibilities to physically remove tars, by means of condensing them and scrubbing the gas. These methods add to the problems rather than abating them, since it creates an additional wastewater stream which also posses a problem for cleaning. Specially designed reactors are needed for abatement of tar containing solution. A better solution should be to separate and recycle tars, which are also being investigated. The existing methods to treat the gas outside the gasifier in a secondary reactor, are by catalytic or by thermal cracking methods. Thermal cracking requires temperatures in the order of 1200 ~ 1400 deg C. One drawback of thermal cracking is that part of the enthalpy of the gas is lost. For catalytic gas cleaning dolomites, olivine, steam-reforming catalysts etc. are the common candidates. This process requires high temperatures (850~900 deg C), which are close to the gasifier exit temperatures. Critical issues are long-term reliability as well as catalyst lifetime [1].

Non-thermal plasma offers a viable alternative to the conventional processes because of its ability to create energetic electrons. Thus reactive species can be generated in any gas environment and thus reactions can be initiated at any gas temperatures. Various options exist to create non- thermal plasma. The present paper, however does not intend to offer comparisons between various such options. Nevertheless, to summarize the option exists to modify the reactor geometry and use a "cheaper" and simple power source or to have a simple reactor and let the power source create the plasma. In the former case, one can use a dielectric barrier discharge reactor (DBD) where the limitation is for long-term usage and its applicability for harsh environments such as gasification gas at higher temperatures and in combination with dusts and corrosive environments. In the later case a complex power source does the job of creating the plasma and no limitation exists on the reactor side. More commonly this is referred as pulsed corona plasma reactors, where a pulse power source feeds high voltage pulses having width in the order of nanoseconds at desired operating frequency. Present available pulse power source can deliver an average power up to 30 kW and further development till 100 kW is underway. Streamer corona plasma created in such a way has been evaluated and average electron energy of the order of 10~12 eV has been reported [2]. With regard to the present application, where high gas temperature coupled with high dust concentration, pulsed corona plasma would be the most ideal plasma technique for gas cleaning.

2. Proof of Concept

Prior to laboratory investigations, demonstrations were carried out to verify the concept of tar removal by use of pulsed corona. These were aimed at two things,

1) Demonstrate the ability of continuous and stable operation of pulsed power source;
2) Verify the idea of tar removal.

Two such studies were carried out where a plasma reactor was placed after the gasifier. In the first set of studies [3], it was shown that heavier tar components were broken down into lighter compounds. To further enhance the operation of such a device, the reactor was operated with a DC bias to function as a precipitator at the same time performs tar removal operation [4]. The later tests further confirmed the result of heavier tar fractions being broken down into lighter tar fractions. Additionally it can remove dust particles as well. Thus a gas cleaning option is studied which can function single handedly.

3. Laboratory investigations

For economic feasibility of such a technology, the obvious question is related to the energy consumption. The demonstration tests were only intended to show that this option is worthwhile for further consideration. Hence detailed laboratory investigations are needed.

Out of the various tar compounds that can be present in the gas stream, naphthalene was chosen as the model tar compound. All further lab data are therefore with naphthalene as the model tar compound. Detailed descriptions of the experimental set-up are provided and the issues related to engineering problems for a high temperature plasma reactor are reported as well [5].

4. Experimental Results and discussion

Tar removal experiments are performed in various gas mixtures relevant to the processing conditions and in synthetic fuel gas as well. Table 1 provides a summary of the experimental conditions. Experimental results at 200 deg C indicated an energy requirement of 400 J/L for naphthalene removal in synthetic fuel gas. Also removal of compounds with side chain (eg. toluene, phenol) required higher energy as compared to naphthalene in the order of 700 ~1000 J/L. Experimentally it was determined, that the reactions for tar removal are initiated by CO_2 decomposition. The energy consumption of the process is determined by the efficiency of utilization of the generated O radicals [6,7].

Table 1 Experimental conditions

Experimental operation: Batch wise			
Temperature = 200 ~ 410 deg C			*Pulse power properties*
Pressure = 1 bar			70 ~ 80 kV Peak Voltage
Biogas composition		*Tar Components*	20 ns rise time, 100 ns pulse width
CO	20 %	Naphthalene	50 ~ 80 MW peak power
CO_2	12 %	Toluene	0.5 ~1.2 J/pulse
H_2	17 %	Phenol	50 pulses per second
CH_4	1 %		80 % efficiency (from Mains to corona)
N_2	Rest		

Combining experiments and calculations, a kinetic scheme for naphthalene removal was realized as indicated in figure 1. Sensitivity analysis on the chemical kinetics resulted in identifying the set of reactions that had strong influence on the process.

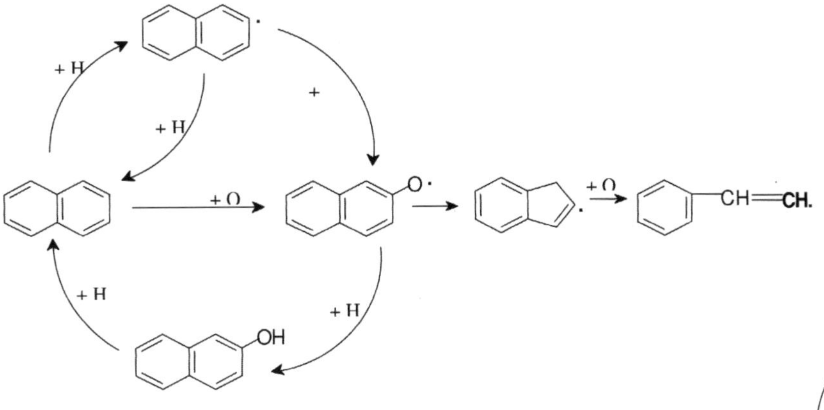

Figure 1 Naphthalene decomposition scheme [8]

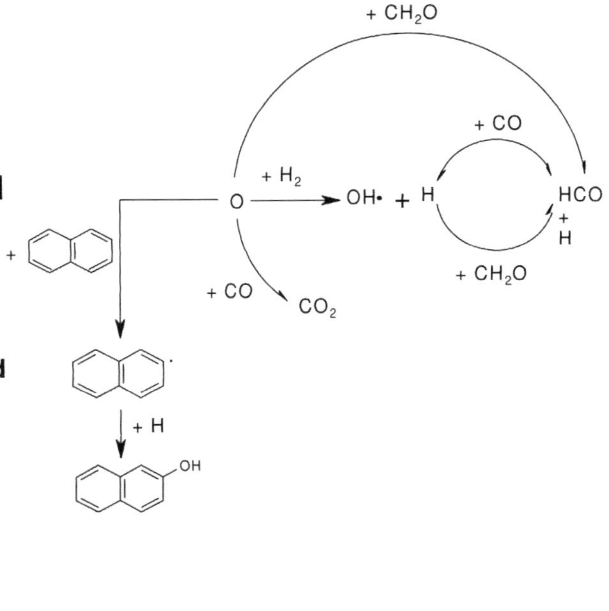

Figure 2: Termination pathways for generated O radicals at 200 °C [8]

The termination pathway or the loss mechanism of the generated O radicals is as indicated in figure 2. The above mechanisms were identified for a temperature of 200 °C [8].

(a) Operation at higher temperatures – dry reforming:

From the above discussion, it is evident that efficiency of the process can be improved by utilizing the radicals effectively. One possible method would be to operate at higher temperatures where the radical kinetics for naphthalene removal would be favored as compared to its loss mechanisms. It was also observed that plasma generation is favored as temperature increases.

$$\beta = \frac{\textit{Energy Transfer to streamer corona (J/pulse)}}{\textit{Energy transfer to matched load (J/pulse)}}$$

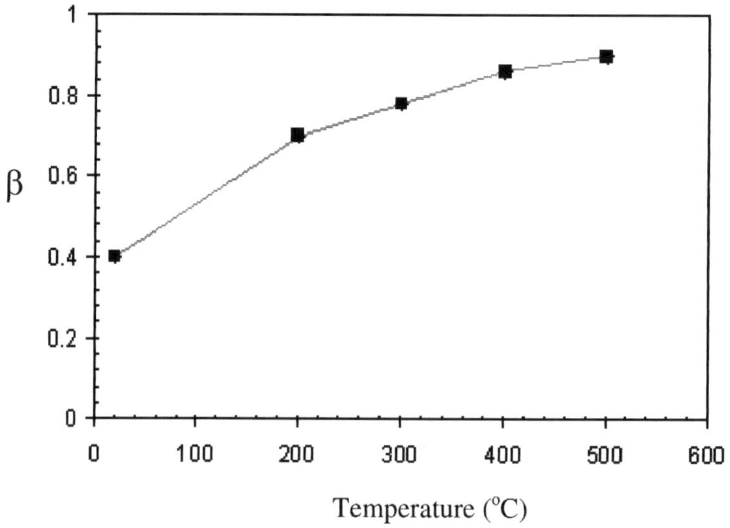

Figure 3 Energy transfer to reactor with increase in temperature

Figure 3 indicates the effect on corona energy transfer with temperature, which indicates an improvement in the electrical efficiency of the device. In addition, a decrease in the energy consumption for naphthalene removal for the dry reforming process was observed experimentally. Figure 4 indicates the energy density for naphthalene removal from a mixture of N_2 and 10% CO_2. The measurements were done with the same initial concentration of 4~5 g/Nm^3.

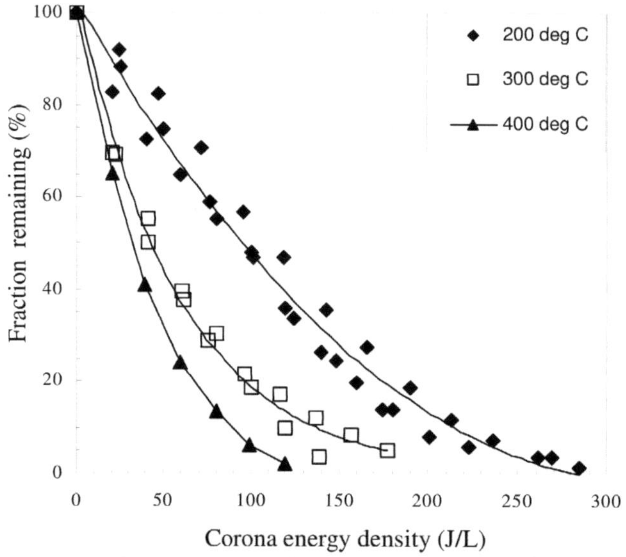

Fig 4: Naphthalene removal from a gas mixture of N_2 + CO_2 (10%) – (Initial conc. 4~5 g/Nm^3)

The decrease in the energy consumption can be attributed to a combination of factors,

a) Favorable kinetics with increase of temperature
b) Uniform plasma generation with increase of temperature
c) Increase of radical yield.

The extent of contribution of each of the above effects cannot be determined from the present available data. However, kinetic calculations done for the case of 200 deg C were extended to the case of higher temperatures. These calculations do take into account the effect of favorable kinetics. But it was observed the favorable situation was account of an increase in the radical yield i.e. higher O radical density. Table 2 indicates the radical yield obtained at higher temperatures.

Table 2: Radical yield at various temperatures

Temperature (deg C)	Radical yield (mol/100eV)
200	1.134
300	1.5687
400	2.429

In reality, the yields indicated above is not a measure from the plasma directly, but it can seen as a measure of effective utilization of the generated species.

(b) *Operation at higher temperatures- synthetic fuel gas*:

Chemical pathways determined at 200 deg C indicated significant loss of radicals by way of competitive reactions. The aim for higher temperature operation was two fold:

a) Gasification gas is obtained at higher temperatures and therefore a gas treatment system should operate at higher temperatures
b) Favorable kinetics for the desired process over the termination pathways.

Experiments were carried out for determining energy requirements in a synthetic fuel gas mixture (composition mentioned in table1). Figure 5 indicates the effect of temperature on the energy consumption for naphthalene removal. The detailed chemical pathways and the mechanism will be studied in the near future.

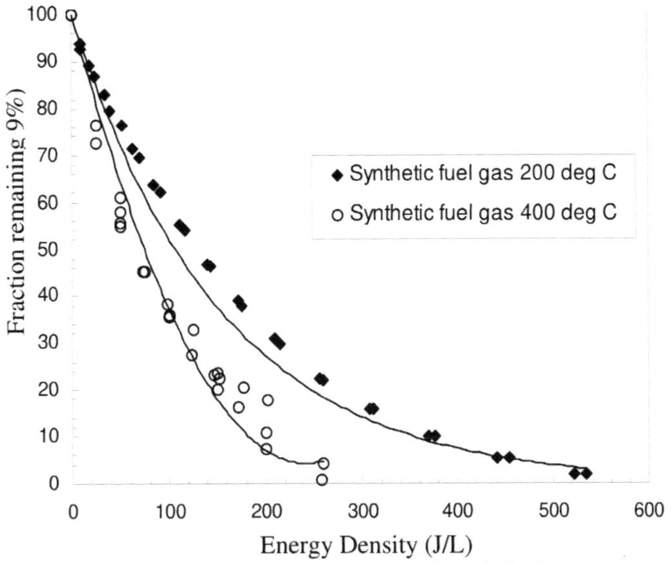

Figure 5: Naphthalene removal from fuel gas (Naphthalene concentration - 4~5 g/Nm3)

5. Conclusions:

For a tar removal system operating a plasma reactor at higher temperature is favorable for improving both the electrical and chemical efficiency.

6. Acknowledgements:

We sincerely thank the financial support from SDE (Dutch association for sustainable technology), ECN (Energy Research Center of the Netherlands) and TDO (Center for sustainable technology, Eindhoven).

References:

[1]. Milne TA, Evans RJ. Biomass gasification "tars": their nature, formation and conversion. NREL, Golden, CO, USA, Report no. NREL/TP-570-25357, 1998.

[2]. Yan.K. Corona plasma Generation, PhD Thesis, Eindhoven University of Technology, 2001.

[3]. B.E.J.M. van Heesch, G.U.A.J.M. Pemen, K. Yan, S.V.B. van Paasen, K.J. Ptasinski, P.A.H.J. Huijbrechts, Pulsed corona tar cracker, IEEE Transactions on Plasma Science 28 (2000) 1571– 1575.

[4]. Nair, S.A.; Pemen, A.J.M., K. Yan; Heesch, E.J.M. van; Ptasinski, K.J; Drinkenburg, A.A.H; Tar removal from biomass derived fuel gas by pulsed corona discharges *Fuel processing technology*, 2003, 84 (1-3), 161-173

[5]. Nair, S.A.; Yan, K; Pemen, A.J.M.; Heesch, E.J.M. van; Ptasinski, K.J; Drinkenburg, A.A.H; A high temperature pulsed corona plasma system for fuel gas cleaning, Paper submitted for *Journal of Electrostatics (Accepted Jan 2004).*

[6]. Nair, S.A.; Pemen, A.J.M.; Yan K, Heesch, E.J.M. van; Ptasinski, K.J; Drinkenburg, A.A.H; Chemical processes in tar removal from biomass derived fuel gas by pulsed corona generation *Plasma chemistry and plasma processing*, 2003,23(4), 665-680.

[7]. Pemen, A.J.M.; Nair, S.A.; Yan, K; Heesch, E.J.M. van; Ptasinski, K.J.; Drinkenburg, A.A.H.; Pulsed Corona discharges for tar removal from biomass derived fuel gas *Plasma and Polymers*, 2003, 8, 209-224

[8]. Nair, S.A.; Yan, K; Pemen, A.J.M.; Heesch, E.J.M. van; Ptasinski, K.J; Drinkenburg, A.A.H; Tar removal from biomass derived fuel gas by pulsed corona discharges – A chemical kinetic study, Paper submitted to *Ind. & Eng. Chem. Res.(Accepted, Jan 2004)*

Geothermal Energy – Opportunities and Challenges

R. Gerald Nix
National Renewable Energy Laboratory
1617 Cole Blvd
Golden, Colorado 80401

What Is Geothermal Energy?

Geothermal energy is the capture and productive use of the heat coming from the depths of the earth. As the slice of the earth below shows, there are very significant temperatures at depth, a considerable heat flux from the earth caused by both normal cooling and by natural radioactive decay, and areas where the flow of heat is enhanced by volcanism and by plate tectonics.

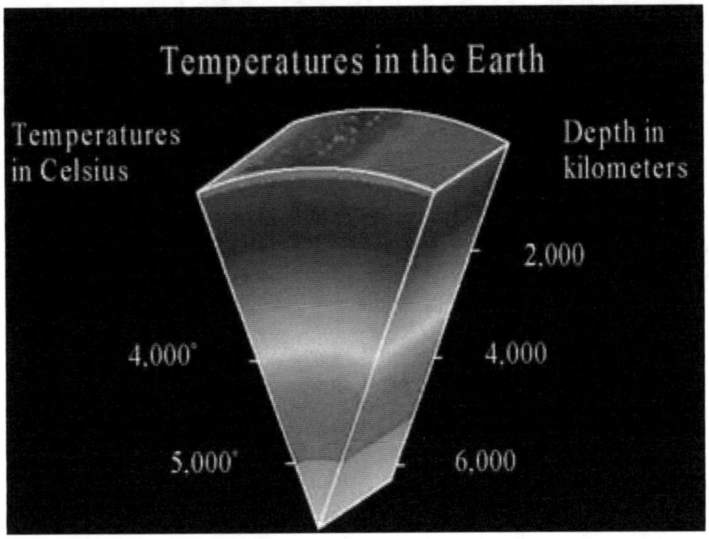

At areas of enhanced heat flux, the earth has been heated to temperatures that are suitable for capture of heat to generate electricity or for direct thermal use. The requirements for a successful hydrothermal geothermal development are significant heat at high temperatures (250-600°F), water to serve as a working fluid to carry the heat, and permeabilities or fractures to allow the water to flow through the rock to convectively transfer the heat.

The heat is carried to the surface for either direct thermal use of for electricity generation. Some of the direct thermal uses are:

- Balneology (hot spring and spa bathing)
- Agriculture (greenhouse and soil warming)
- Aquaculture (fish, prawn, and alligator farming)
- Industrial Uses (product drying and warming)
- Residential and District Heating

Power Plants involve the extraction of the heated water or steam from the ground and direct use in a power conversion cycle, or use as a heat source to vaporize a secondary working fluid (e.g. isobutane) in a binary conversion cycle. The vaporized fluid drives a turbine that is connected to a generator to produce electricity. A generic cycle is shown below:

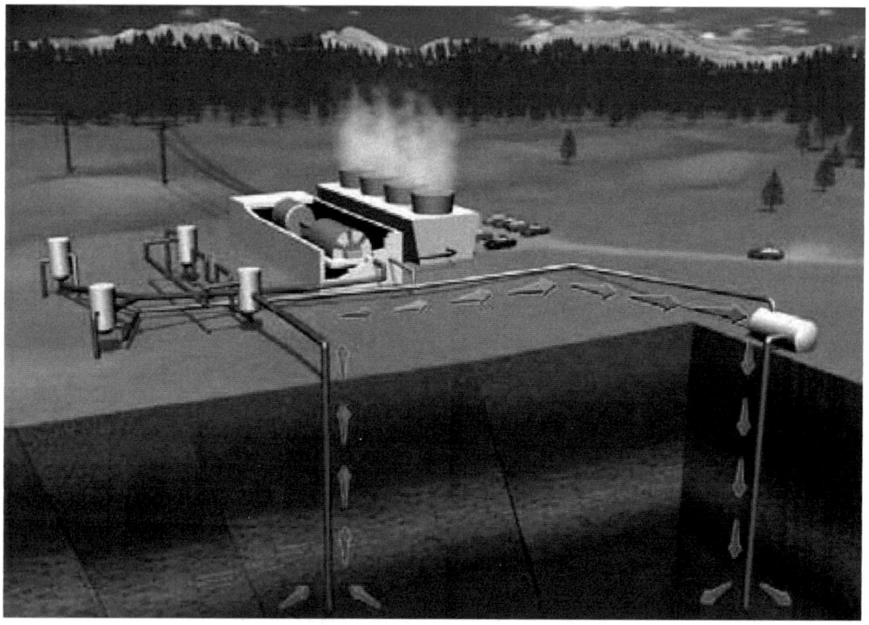

What is the Status of Geothermal Energy?

Geothermal energy is base load energy, available continuously for a sustained period – in fact with proper design, the life-limiting feature is the equipment and is not the geothermal reservoir. Geothermal technology has the following characteristics:

- Well developed and practiced technology
- 2200MW(e) installed in the U.S. and 8000MW(e) worldwide
- Large plants about 5-7 cents/kW-h, small plants about 7-9 cents/kW-h
- Research and development underway to reduce risks and to reduce costs
- Poised for expansion with increasingly favorable investment climate
- Extensive use for direct thermal applications

What Are the Challenges to Advancing Geothermal Energy?

Challenges include:

- **Expanding the Hydrothermal Resource Via Better Exploration Techniques**

The United States Geological Survey published Circular 790 in 1978 stating that the hydrothermal resource in the U.S. is known, discovered resources equivalent to 23,000MW(e), with additional undiscovered hydrothermal resources of 4 to 5 times that amount. Although there is significant question of the validity of the USGS estimates, it appears that we have only developed 2 to 3% of the hydrothermal resources within the U.S. Where are the resources? It is more difficult to find a developable hydrothermal system than it is to find oil or gas!

New exploration techniques are required, especially those suitable for discovering where resources are under the surface when there is no surface expression of the resource's existence. The techniques include enhancements of time-honored geological and geophysical techniques, including seismic, magnetic and resistive techniques. Newer tools include GIS (geographical information systems) and satellite imaging. The GIS allows easy overlaying of a significant amount and types of data to develop indicative trends. Satellite imaging allows use of innovative techniques such as looking for effects of otherwise difficult-to-detect emissions or heat, such as heat stressed vegetation. Exploration for new geothermal systems is an exciting area of research – one that is necessary for geothermal energy to become a major energy source in the U.S.

- **Better Drilling Techniques at Lower Costs**

Regardless of the exploration technique, it is still necessary to drill holes in the earth to confirm, characterize and to utilize the geothermal resource. Significant research is underway to develop better conventional drilling techniques and to develop innovative drilling techniques. The ultimate is "smart" drilling for conventional rotary drilling, where a computer with appropriate software and data input controls the drilling in real time, making the adjustments necessary to handle changing rock types and to allow for steering to the most productive areas. The concept includes using the signal from the drill bit to "look ahead" and geochemical sampling while drilling as information to steer the drill. Passive and active dampening devices would reduce the damaging impulsive forces on the drill, all of which would increase the rate of penetration, time-in-hole and would reduce costs.

Development of such a system requires high temperature electronics capable of withstanding temperatures greater than 600°F, with rapid transmission to the surface to enable real time control. New software and imaging techniques would be required. This is an opportunity for adaptive control schemes.

Another opportunity is for innovative drilling techniques, something other than a crush and grind mechanism for advancing through rock. This can include concepts ranging from mechanical impact, to laser boring, to sonic boring, to spallation jets, to whatever an engineer can imagine and cost-effectively reduce to practice.

- **Reservoir Engineering to Provide Enhanced Geothermal Systems**

While the hydrothermal resource estimates in USGS Circular 790 are significant, there exists a virtually inexhaustible geothermal resource found across the entire U.S. – a resource capable of supplying the energy needed to generate the entire U.S. electricity supply for hundreds of thousands of years. At depths of 2-3 miles, there is significant hot rock across the entire U.S., usually with presence of very little water and with unfractured formations. While most of the geothermal development has occurred in the Western U.S. deeper drilling can access hot rock across the U.S. The geology found in the Western U.S. is considerably younger than that in the Eastern U.S. resulting in the requirement to go to greater depths in the Eastern U.S. to find rock suitable for "heat mining". The other requirement is the capability to cost-effectively and controllably form underground reservoirs via fracturing techniques. Water can then be injected to act as a working fluid that can be extracted to drive conversion plants to generate electricity. Not only does this technique allow development across the U.S., but also it allows for an orderly development path, starting with an extension of existing hydrothermal systems and later development of systems where the rock is unfractured and the amount of water present is limited.

- **Reducing Costs for Electric Generation**

Chemical engineers are well suited to answer the challenges of improving power conversion cycles. There are many variations of the Rankine cycle that promise to increase efficiency, such as the Kalina cycle that uses an ammonia-water working fluid. The Kalina cycle is a typical mixed working fluid cycle that allows a closer approach to equilibrium than can be obtained with a pure working fluid, with a higher conversion efficiency because of reduced entropy generation. The power plant is the most immediate opportunity for reduction in the cost of geothermally generated electricity.

A typical opportunity is in the rejection of heat in the working fluid condensers. It is less expensive to reject heat to water, but if cooling water is unavailable, then it is necessary to used finned air-cooled condensers. How do you significantly enhance the performance of air-cooled condensers, especially through a significant increase in the airside heat transfer coefficient? The major limitation in current airside heat transfer rates is because the air boundary layer on the fin becomes stagnant – any geometry change that renews the boundary layer on the fin, while not causing significant additional parasitic pressure drop, will result in enhanced performance, a smaller and hopefully less-expensive heat exchanger, and reduced cost of electricity. Methods of boundary layer renewal include use of small vortex generators or flow directors on the fin surface.

Another opportunity is in materials of construction. Many geothermal brines are aggressive, requiring higher alloys – what about coated carbon steel? A coating that has been commercialized is doped polyphenylene sulfide, with dopants including chopped Kevlar and carbon fibers, and Teflon. This coating has a thermal resistance of about 0.001 ft^2xhrxF/BTU, with exceptional corrosion resistance and reduced fouling. It has been commercialized as Curralon.

Another opportunity is to generate an additional revenue stream via recovery of minerals and metals from the geothermal brine. This is essentially "solution mining by nature", with recovery via physical processes, hydrometallurgical techniques and electrochemical techniques.

What's in Geothermal Energy for Chemical Engineers?

Geothermal energy is a chemical engineer's dream! There are many opportunities for application of the tools and talents of the chemical engineer including: fluid flow, conversion cycle engineering, heat transfer, environmental engineering, materials engineering, hydrometallurgical techniques, and many others. With declining fossil resources, there are many opportunities for development of geothermal energy for both electricity generation and for direct thermal use, and along with it opportunities for chemical engineers to advance the technology and to find exciting employment.

WASTE-HEAT POWERED AMMONIA ABSORPTION REFRIGERATION IN PROCESS INDUSTRIES

Donald C. Erickson and C.B. Panchal

Energy Concepts Company, LLC
627 Ridgely Avenue
Annapolis, MD 21401

Key Words: ammonia absorption cycle, waste heat utilization, refinery, petrochemical

Prepared for Presentation at the 2004 National Meeting, New Orleans, LA, April 2004

Unpublished

WASTE-HEAT POWERED AMMONIA ABSORPTION REFRIGERATION IN PROCESS INDUSTRIES

Donald C. Erickson and C.B. Panchal
Energy Concepts Company, LLC

627 Ridgely Avenue
Annapolis, MD 21401

Extended Abstract

This tutorial will provide an overview of ammonia absorption refrigeration cycles and discuss specific applications in refining and chemical industries.

Ammonia absorption refrigeration (AAR) converts waste heat to high-value refrigeration. Refineries are the largest and most concentrated sources of waste heat among the process industries. Fired heater stack gas at 450°F - 600°F and process streams at 250°F - 400°F are two main sources of waste heat. Many other chemical processes have similar sources of waste heat. The AAR uses waste heat to produce chilling down to minus 50°F. That refrigeration can be used to improve various process industry operations:

- Recovery of LPG from reforming and FCC process units and eliminate flaring
- Debottlenecking FCC and coker compressors, and crude vacuum column
- Enrichment of hydrogen from tail gases
- Traditional refinery refrigeration in alkylation and lube-oil de-waxing processes
- Traditional petrochemical refrigeration in ethylene, NGL, etc.
- Enhancement of separation processes by sub-ambient cooling

The AAR can improve the energy efficiency of conventional and emerging separation processes. Examples include: a) AAR integration with pressure-swing adsorption (PSA) for gas separations, particularly for enriching hydrogen; b) AAR integration with membrane systems such as per-vaporization (PerVap) membranes; and c) sub-ambient distillation processes.

Ammonia absorption technology can also be used for co-production of electric power and refrigeration. The dual-function absorption cycle produces power and refrigeration interchangeably using the same equipment. Ammonia absorption cycles provide the same exceptional levels of efficiency and economy in converting waste heat into power as they do for refrigeration.

MATERIALS—KEY TO HYDROGEN ECONOMY

XiaoFeng Xie
Institute of Nuclear and New Energy Tech.
Tsinghua University
Beijing 100084
CHINA

V. K. Mathur
Department of Chemical Engineering
College of Engr. & Physical Sci.
University of New Hampshire
Durham, NH 03824, U.S.A.

Toby Hamblin
Lydall Filtration/Separation, Inc.
Rochester, NH 03866, U.S.A.

There is a general awareness amongst the scientific and industrial communities that the cost of fuel cell materials must be reduced so that hydrogen economy can be realized. Some of these materials are membranes, electrodes with associated catalysts, bipolar plates and electrolytes. Other critical needs are in the area of hydrogen production, hydrogen storage and methanol as a hydrogen provider. This paper deals with materials particularly in the area of membranes and gas diffusion layers (GDLs).

The perfluorinated polymer electrolyte membranes such as Nafion TM (DuPont), Aciplex (Ashahi Chemicals Co.), Flemin (Ashahi Glass Co.) and Dow (Dow Chemical Co.) have been extensively used as polymer electrolytes for fuel cells because of their stable electrochemical and mechanical properties. However, the perfluorinated polymer electrolyte membrane is not appropriate for use in DMFC as it allows methanol crossover to cathode. Its penetrability is up to 40%. To overcome this problem, the development of new proton-conducting polymer electrolytes is necessary for practical applications.

China has several projects where the use of PEMFC is anticipated as the energy source, particularly in the area of transportation. Direct methanol fuel cell (DMFC) using a polymer electrolyte membrane is a promising candidate for the applications in lab tops, cell phones, and other portable devices because it does not require any fuel processing equipment. However, current polymer electrolyte membranes suffer from several disadvantages such as high cost and methanol crossover. A challenging problem in the development of a DMFC is to find a proton exchange membrane that can effectively enhance power density while not allowing methanol crossover.

Membrane Preparation

One of the authors of this paper has worked towards the synthesis of block copolymers of monomers with different chemical nature. This work consists of first synthesizing a modified polybenzimidazole monomer (mPBI) and then preparing block copolymers for making proton exchange membrane using living ionic polymerization techniques. The conductivity of proton exchange membrane is developed by introducing $-SO_3$ groups in monomers. The proton conductivity is measured to be 3.12×10^{3} Scm^{-1}. The membrane is mechanically very stable.

The measured permeation of the membranes is given in Fig 1. Three samples of membranes mPBI-1, mPBI-2, and mPBI-3 containing various amounts of SiO_2 are investigated. The methanol permeations of the membranes from mPBI are all lower than that for Nafion117. The mPBI membranes have a Heteropolyacids backbone structure causing completely different permeation rates and other properties relative to those of Nafion. Moreover the membranes from mPBI are also thinner than Nafion117, the molar flux of methanol through the materials is significantly lower and thus the expected benefit of reduced methanol crossover to cathode can be realized.

Gas Diffusion Layer(GDL)

Another component that it is vital to the performance and durability of the fuel cell is the gas diffusion layer or media or electrode (GDL or GDM or GDE).The GDL consists of a carbon substrate that can either be woven or non-woven, onto which is coated a carbon catalyst layer, usually consisting of a mixture of carbon black and PTFE – controlling hydrophobicity while ensuring conductivity.

The various substrates used to produce GDLs result in very different physical characteristics and therefore in the performance of the fuel cell. Probably the most well-known GDL substrate used during the development of the PEM fuel cell was that produced by Toray of Japan. This substrate is produced by pyrolysing a non-woven carbon-fiber sheet which gives a sheet very good conductive properties and low elongation but limited resistance to brittle fracture. In general, the carbon/PTFE layer is applied by the end-user with the amount depending on the hydrophobicity required.

Another approach to GDL manufacture is to use a woven carbon cloth as the substrate, the best known example being Elat® from Etek, a US company which is part of the Italian de Nora group. The cloth is produced using highly conductive fibers which are then "loaded" with carbon black. The carbon/PTFE coating can be applied either by the substrate manufacturer or the end-user. Cloth based GDLs have the advantage of being very flexible and durable but their tendency to stretch can make automated assembly of a stack rather difficult. Woven substrates also tend to be more expensive than similar non-woven sheets.

A rather different approach is being taken by Lydall with its LyFlex® GDL where the substrate is again a non-woven carbon sheet but does not need to be pyrolysed and remains very flexible. In addition, as there are no high-temperature post-processes, the carbon/PTFE can be integrated directly during the substrate manufacturing process – eliminating a costly production step. The advantage of such an approach is not only the low cost of the GDL itself but its ability to be supplied in roll-form for use in industrial stack-assembly processes. This type of material is, however, inherently more compressible than a pyrolysed sheet and will experience structural failure if too high a pressure is applied.

In a study at Tsinghua University, a membrane electrode assembly (MEA) is put together using the modified PBI membrane and Lydall's LyFlex GDL paper. Three samples of membranes mPBI-1, mPBI-2, and mPBI-3 are investigated. The anode catalyst is 35 wt% Pt, 15wt% Ru. The anode catalyst is 15wt% Pt on carbon (Johnson Matthey, fuel cell grade). The catalysts are coated on GDLs surface facing the membrane. Methanol is used as a fuel.

Determination of current-voltage curves

Current-voltage curves are determined with Potentiostat Galvanostat Model 263A (Princeton Applied Research). Impedance spectra are obtained at frequencies between 1 kHz and 3 mHz with 10-15 steps per decade. Tests on the DMFC are performed with an active area of 5 cm^2. Aqueous methanol solution is fed by mass flow pump at 4ml/min. Air is fed at a flow rate of 1.5 l/min at a pressure of 2 kg/cm^2. MEAs using three mPBI membrane samples are used for studying the performance of DMFC at various experimental conditions.

Fig.2 shows a comparison of the voltage and current data for MEAs made from Nafion 117 and mPBI membranes under the same experimental condition. It is evident that MEAs made from mPBI with Lydall's GDL show higher power density than those from Nafion membrane. There is a slight improvement in cell voltage, approximately 15mV. The data indicate that the electrical resistance of MEA with the mPBI membranes and Lydall's GDL is lower than with Nafion117. However, on continuous operation of MEA made with the mPBI membranes the performance degrades rapidly. On removing the MEA from the cell de-lamination between the catalyst and membrane layers is observed.

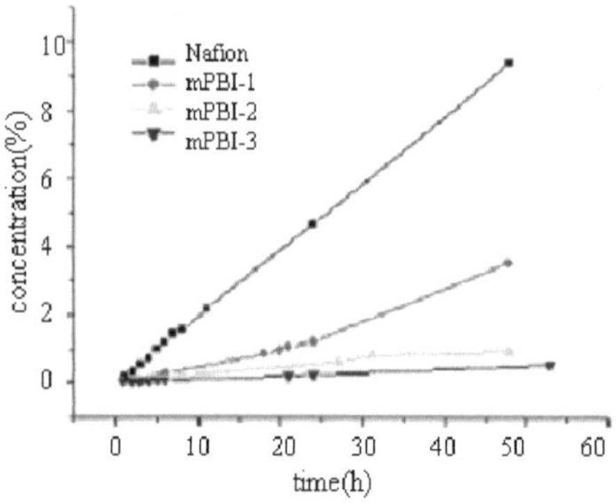

Fig.1 Methanol permeation of membranes

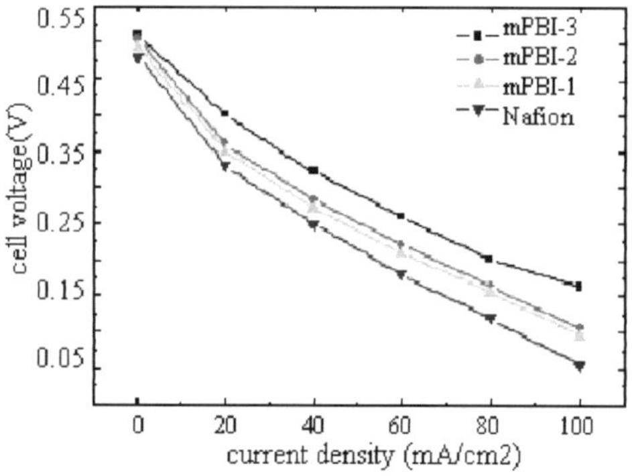

Fig.2 Comparison of the voltage and current of MEAs made with Nafion and mPBI

Conclusions

This study indicates that MEAs made from mPBIs and Lydall's LyFlex GDL have potential application in the DMFC. The power performance of MEAs assembled from these materials is as good as those made with Nafion membrane with lower methanol crossover. However, further research work is needed to find ways to prevent delamination of catalyst layer from the membrane surface to produce high performance DMFC.

Reference

Xiaofeng Xie, Zongqiang Mao, JingMing Xu, "A Hybride Membrane of Modified Polybenzimidazole and Heteropoly Acid for Direct Methanol Fuel Cell", Power Con 2002, IEEE/PES-CSEE International Conference On Power System Technology, Vol 1-4, 169-172, Kunming, China

V.I. Basura 1, C. Chuy, P.D. Beattie 1, S. Holdcroft, "Effect of equivalent weight on electrochemical mass transport properties of oxygen in proton exchange membranes based on sulfonated a,b,b-trifluorostyrene and sulfonated styrene-(ethylene-butylene)-styrene triblock copolymers", Journal of Electroanalytical Chemistry 501 (2001) 77–88

Improvement of Cathode Materials for Polymer Electrolyte Fuel Cell

Ken-ichiro Ota, Akimitsu Ishihara, Kunchan Lee, Toshihiko Nakagawa, Shigemori Mitsushima, and Nobuyuki Kamiya*

Department of Energy and Safety Engineering, Yokohama National University
79-5 Tokiwadai, Hodogaya-ku, Yokohama-shi, Kanagawa 240-8501, Japan
** Core Research for Evolutional Science and Technology (CREST), Japan Science and Technology Corporation (JST)*

2004 Prepared for Presentation at 2004 Spring National Meeting/ April 26, 2004/ Alternate Energy Systems I

Abstract: A large cathode overvoltage of polymer electrolyte fuel cell decreases the energy and exergy efficiency. An improvement of cathode materials should be needed to spread the fuel cell systems. An interesting carbon material, Vapor Grown Carbon Fiber (VGCF™), has been tried to apply to carbon support for the purpose of the improvement of the catalytic activity of oxygen reduction. The WC based material has been also tested as the substitutive material of Pt. Ta added WC showed the improved stability and the catalytic activity for oxygen reduction.

1. Introduction

A large overvoltage of the oxygen reduction reaction (ORR) at the cathode of polymer electrolyte fuel cell is the important problem in order to obtain a higher efficiency, and it should be reduced for practical commercialization. We have tried to improve the cathode materials by mainly two points of view as follows: 1. Application of new materials such as vapor grown carbon fiber to carbon support, 2. Evaluation of non-platinum material as a substitutive catalyst of Pt.

"1. Application of new materials such as vapor grown carbon fiber to carbon support" was performed to increase the catalytic activity of dispersed platinum. Since Vapor Grown Carbon Fiber (VGCF™) has small surface area and high degree of crystallization, platinum was not highly dispersed on its surface to increase the particle size. A high-speed mill was used to cut off the fibers and roughen the surface. The dependence of operating time of the mill (named the cutting time) and the effect of hydrophilic treatment of carbon fiber were examined.

"2. Evaluation of non-platinum material as a substitutive catalyst of Pt" was examined for researching various materials as a non-Pt catalyst for ORR. The estimated amount of the Pt deposits was too little to make many fuel cell systems, especially fuel cell vehicles. Thus, for the popularity of the fuel cell systems in the future, the development of an alternative Pt catalyst is required. Tungsten carbide was used to investigate the applicability to a non-platinum catalyst because of its platinum-like electronic structure [1] and high electric conductivity. Levy et al. reported that WC showed a similar chemisorption of hydrogen property as Pt [2]. Many trials have been performed to apply WC to a fuel cell anode [3-4]. However, since the cathode of the PEFC is in a severe corrosive environment such as in an acidic and oxidizing atmosphere, the carbide is not stable and little studies have been performed for its application as the cathode. We added valve metals, such as tantalum, to improve the stability of the carbides.

2. Experimental

2.1 Electrochemical measurement

The electrocatalytic activity for ORR in acid electrolyte was evaluated by electrochemical measurements. A conventional three-electrode cell was utilized, and when the sulfuric acid was used as the electrolyte, the reversible hydrogen electrode (RHE) was the reference in the same solution. For the study of carbide, a solid-state cell which used Nafion®117 as the electrolyte was applied, and the dynamic hydrogen electrode (DHE) was used as a reference electrode. Cyclic voltammetry was performed to characterize the prepared catalysts, and slow scan voltammetry was performed under oxygen atmosphere to examine the catalytic activity of ORR.

2.2 Catalyst preparation

2.2.1 Application of new materials such as vapor grown carbon fiber to carbon support

VGCF™ was cut off a certain time named cutting time by high-speed mill (Wonder blender; 25,000rpm). As a hydrophilic treatment, carbon fibers were immersed 5 hours in heated 60% nitric acid solution. VGCF performed hydrophilic treatment was expressed as VGCF™-H. The supported Pt catalysts were prepared according to conventional impregnation methods in the literature [5]. After the carbons of 0.45g were added to 50 cm^3 distilled water by ultrasonic treatment and stirred for several tens of minutes to produce homogeneous slurry, 0.172g Na_2CO_3 was added to the slurry. The temperature of the slurry was raised to 80 °C. A solution of 0.303 g Pt as chloroplatinic acid, in 50 cm^3 distilled water, was added to the slurry. The temperature of the slurry was maintained at 80 °C for two hours. An excess formaldehyde solution was added and the slurry was stirred for an hour and then filtered to collect the Pt supported carbon catalysts and washed with distilled water to remove the soluble chloride species. The catalyst was dried at 120 °C for 6 hours in air. The material was then heated to a temperature of 400 °C in a flowing argon atmosphere for 2 hours. This produced a catalyst with a composition of 20wt % Pt. Pt-supported carbon catalysts were observed by TEM to measure the mean particle size of the Pt.

2.2.2 Evaluation of non-platinum materials as a substitutive material of Pt

The WC or the WC to which tantalum was added was deposited on a substrate (glassy carbon rod: diameter=5mm) by RF sputtering. A WC target was used to obtain both the WC and WC-containing Ta film electrodes. In order to prepare the WC-containing Ta, metal tantalum pieces (thickness=0.1mm, Area=1cm^2) were put on the WC target and sputtered together. The composition ratio of W and Ta was controlled by the ratio of the target surface area. The composition of the sputter-deposited carbide was determined by energy dispersive X-ray analysis (EDX).

3. Results and discussion

3.1 Application of new materials such as vapor grown carbon fiber to carbon support

Figure 1 shows the TEM photograph of Pt supported VGCF^{TM}-H catalyst with cutting time of 20minutes. Mean Pt particle size of Pt/VGCF^{TM}-H with no cut was about 16nm. Whereas, that of Pt/VGCF^{TM}-H with cutting time of 20min. was ca. 2-3nm as shown in Fig.1. This meant that platinum complex ion was easier to adsorb on the surface of cut VGCF^{TM} and the points of nuclei formation increased because the surface of VGCF^{TM} was roughened. Therefore, platinum was highly dispersed as the same as the Pt/Vulcan XC-72R used commonly.

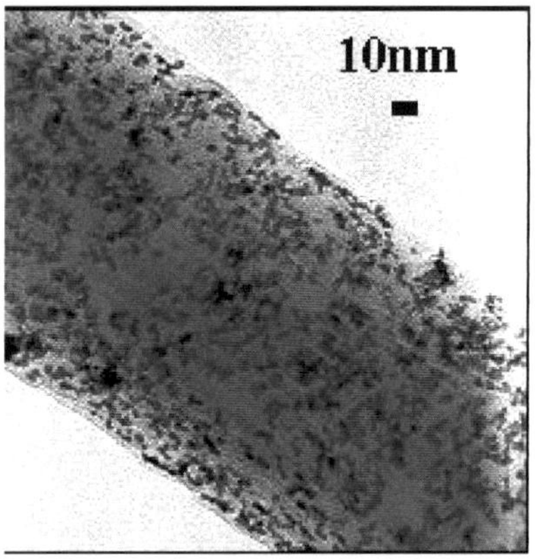

Fig.1 TEM photograph of 20wt% Pt/VGCF^{TM} -H catalyst. Cutting time: 20min.

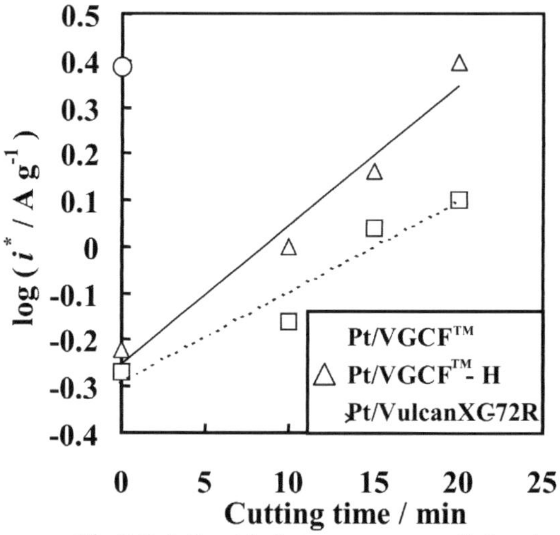

Fig.2 Relationship between mass activity at 0.9V vs. RHE and cutting time.

Figure 2 shows the relationship between mass activity at 0.9V vs. RHE of oxygen reduction and cutting time. Mass activity increased with increasing the cutting time. This was because the roughness of the surface increased with increasing the cutting time. Below 20minutes of cutting time, the hydrophilic treatment made the mass activity increase. This suggested that the functional group of carbon surface affected the dispersion of platinum. Pt/VGCF^{TM}-H with cutting time of 20minutes had similar mass activity of Pt/Vulcan XC-72R which is used as a commercial catalyst.

3.2 Evaluation of non-platinum materials as a substitutive material of Pt

Figure 3 shows the steady-state cyclic voltammograms of WC and the WC+Ta on the Nafion® under N_2 atmosphere. A current density was expressed by geometric area. A large anodic current in the WC was observed above 0.6V vs. DHE. WC violently corroded, that is, both W and C were oxidized to WO_3 and CO_2, respectively. [6]

$$WC + 5H_2O = WO_3 + CO_2 + 10H^+ + 10e^- \quad (1)$$

Tantalum, known as a valve metal, was added to the WC in order to increase the corrosion resistance. In case of WC containing Ta, little anodic current was observed below 1.0V, as shown in Fig.3. With addition of Ta, the

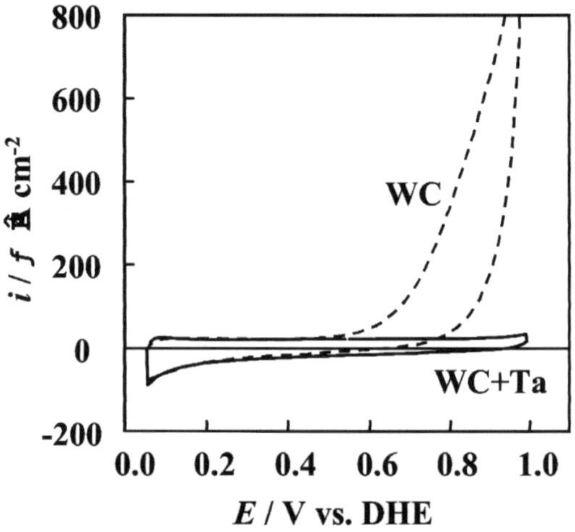

Fig.3 Cyclic voltammogram of WC and Ta added WC. Nafion®117, Scan rate=100mV/s, 30

oxidation of W and C ceased. Ta formed tantalum oxide (Ta_2O_5) of which the corrosion resistance is high in acid and under an oxidizing atmosphere. Though W and C on the surface were dissolved just after contact with Nafion®, Ta_2O_5 was rapidly formed in the electrode surface afterwards, and the corrosion resistance increased.

In order to investigate the catalytic activity for ORR and stability, slow scan voltammetry was carried out under N_2 and O_2 atmospheres. Figure 4 shows the results for the WC and the WC+Ta under each gas atmosphere. The rest potential of the WC electrode was low, i.e., ca. 0.45V, and the anodic current was observed at the higher potential. However, below 0.4V, the reduction current of WC under O_2 was greater than that under N_2. This suggested that WC might have some catalytic activity for ORR. On the other hand, the WC+Ta had little anodic current up to 1.0V under N_2 atmosphere, because of the formation of thin oxide film on surface. In case of WC+Ta, a reduction current was observed below about 0.8V, and it was proven to have the catalytic activity of ORR. Because Ta_2O_5 had little catalytic activity, it was considered that WC which has some catalytic activity for ORR existed on the surface and/or sub-surface. Figure 5 shows the XPS spectra for the W 4f and the Ta 4f after the electrochemical measurements. In case of WC, the XPS peaks for carbidic tungsten disappeared and the peak of oxidic tungsten appeared after the electrochemical measurements. On the other hand, on the surface of the WC+Ta, the carbidic tungsten as well as the oxidic tungsten was observed after the electrochemical measurements. The electrocatalytic activity observed in the WC+Ta was considered to be due to the carbidic tungsten that exists on the surface and/or sub-surface.

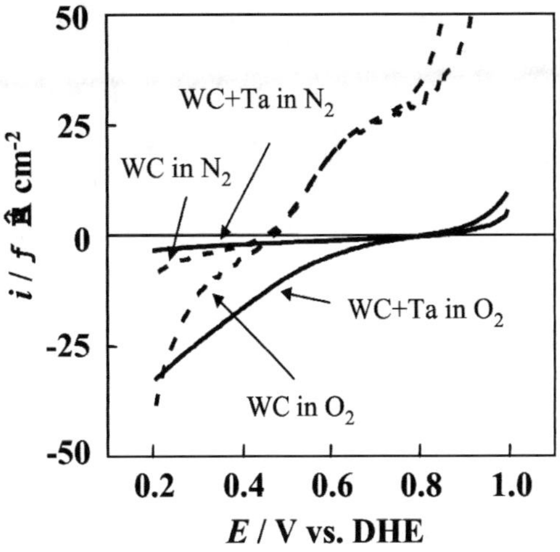

Fig.4 Potential-current curve of WC+Ta on Nafion®117. Scan rate = 5mV/s, 30 •

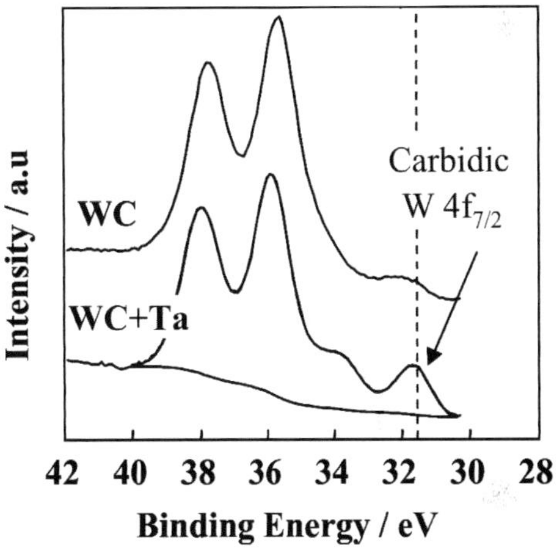

Fig.5 XPS spectra for W 4f after electrochemical measurements.

4. Conclusion

We have tried to improve the cathode materials by mainly two points of view as follows. A new carbon material, Vapor Grown Carbon Fiber (VGCF™), has been tried to apply to carbon support for the purpose of the improvement of the catalytic activity of oxygen reduction. Platinum was highly dispersed with the particle size of 2-3 nm on the VGCF™ which was roughened the surface. The WC based material has been also tested as the substitutive material of Pt. Ta added WC showed the improved stability and the catalytic activity for oxygen reduction. The electrocatalytic activity observed in the WC+Ta was considered to be due to the carbidic tungsten that exists on the surface and/or sub-surface.

Acknowledgment

The authors thank Showa Denko Co. Ltd for supply the Vapor Grown Carbon Fiber (VGCF™).

References

[1] L. H. Bennett, J. R. Cuthill, A. J. McAlister, N. E. Erickson, *Science*, **184**, 563 (1974).
[2] R. B. Levy, M. Boudart, *Science*, **181**, 547 (1973).
[3] D. R. McIntyre, G. T. Burstein, A. Vossen, *J. Power Sources*, **107**, 67 (2002).
[4] R. D. Armstrong, A. F. Douglas, and D. E. Keene, *J. Electrochem. Soc.*, **118**, 568 (1971).
[5] L. Keck, J. Buchanan, and G. A. Hards, US Patent, 5,068,161 (1991).
[6] J. D. Voorhies, *J. Electrochem. Soc.*, **119**, 219 (1972).

The effects of air purge on the degradation of PEM fuel cells during startup and shutdown procedures

Paul T. Yu, Shyam Kocha, Lesley Paine, Wenbin Gu, Frederick T. Wagner

Fuel Cell Activities
General Motors Co.
Honeoye Falls, NY 14472-0603

Key words: Fuel cells, Short stack, Startup, Shutdown, Degradation

Prepared for presentations at the 2004 Annual Meeting, New Orleans, LA, April 25-29

Abstract

Uncontrolled stack shutdown and startup can result in stack performance decay due to entry of air into the hydrogen side of the stack. This phenomenon was studied both on 15-cell full-active-area short-stacks and on small-scale (50-cm^2) single cells by means of 4-way valves installed in the anode feed-line directly upstream of the fuel cell. In both cases, very significant *cathode* degradation was observed when air was purged repeatedly through the *anode* gas compartment that was filled with hydrogen. Scanning Electron Microscopy (SEM) pictures of membrane/electrode assemblies (MEAs) subjected to this treatment indicated that carbon corrosion on the cathode is likely the major reason for the observed cell voltage degradation, in accordance with a proposed (complex) deactivation mechanism. The effects of anode air purge flow rate were investigated on the 50-cm^2 small-scale cells. The degradation rate was found to be proportional to the residence time of the hydrogen/air front within the stack. Even front residence times less than 0.1 second do not ensure the attainment of strict automotive durability targets. Further improvements in materials and operating procedures are needed to fully mitigate shutdown/startup losses.

Introduction

Recently, polymer electrolyte membrane fuel cells (PEMFCs) have been widely studied for various applications due to their high power density, efficiency and prospects of environmental friendliness. These reasons have drawn from automakers a lot of attention towards developing fuel cell vehicles for replacing the conventional internal combustion engine (ICE) in the main drive train. In recent years, PEMFC vehicles showing progress have been demonstrated and tested on the road. However, PEMFCs still face many challenges in many areas such as humidification, capability to withstand freezing conditions, reliability, durability as well as cost. These challenges need to be overcome to allow PEMFCs vehicles to be offered on the mass market.

Automotive durability requires ~5,000 hours operation time and ~30,000 startup/shutdown cycles without having any significant voltage loss. With these strict vehicle durability requirements, each step of operation and each part of the stack could be a reason for PEMFC degradation. Degradation sources could come from: water management, impurities, reformate, materials, operation conditions and uniformity during operations [1]. Furthermore, these degradation sources could link together yielding a very complex degradation mechanism, which is not easily de-convoluted into its individual effects.

In the past, a relatively small number of durability studies have been published due to the high cost associated with fuel cell lifetime tests and the significant amount of time needed to implement such tests in real time. Also, most durability studies to date have dealt with steady state performance decay. Little published effort has gone towards stack startup and shutdown with the exception of a few patent applications [2,3]. Startup/shutdown on phosphoric acid fuel cell (PAFC) units for electric utilities has been studied by Landsman et. al. [4]. They described that the electrode on both anode and cathode could experience Pt dissolution and carbon corrosion if the electrodes were exposed to air when the stack was hot and held at open circuit (zero load). It has proven necessary to blanket the stack with N_2 when it starts up and shuts down. However, in the automobile application, on-board N_2 is not available. The operation temperature of the PEMFC is around 80 °C, significantly lower than the PAFC operation temperature of 205 °C. To provide stable gas compositions during storage, it would be desirable to purge both side of an automobile stack with air prior to shutdown. This paper quantifies the damage done to stacks by such purging with air. In this study, air purges were investigated both on 15-cell full-active-area short-stacks and on small-scale (50-cm^2) single cells.

Experimental

The effect of air purge through the anode on the degradation of a PEMFC during startup/shutdown procedures was studied by using a 4-way valve installed in the anode feed-line directly upstream of the fuel cell. Figure 1 shows a schematic diagram illustrating the gas flow direction and the means of simulating fuel cell startup and shutdown. Both 50 cm^2 small-scale single cell and 15-cell full active area short stack have been tested in apparatuses described by Figure 1. A 4-way valve is the key experimental element. It allows anode gas switching back and forth from hydrogen to air in order to simulate shutdown and startup. In the shutdown mode, the anode is initially filled with hydrogen and is then purged out by air. On the other hand, in the startup mode, the anode is initially filled with air and is then purged with hydrogen.

The operating parameters and procedures for the startup/shutdown cycle test are described as follows: The 50 cm^2 PEM fuel cell or full active area PEM fuel cell was assembled and connected to the test system as shown in Figure 1. Then, it was pre-conditioned (cycled among 0.6V, 0.85V and open circuit) and fully humidified at 80 °C, 150 kPa. Once the cell was ready, the anode was fed 100% fully humidified ultra pure H_2 while the cathode was fed 100% fully humidified air with 50% utilization feed flow rate on both anode and cathode. The cell or stack voltage was measured at 0.05, 0.2 or 0.8 A/cm^2 before and after each pair of startup and shutdown cycles in order to examine cell degradation due to startup/shutdown. Once the cell voltage under load was measured initially, the load was turned off for 2 minutes in order to stabilize purge gas (air or N_2) flow. Meanwhile, both anode and cathode were still fed with humidified H_2 and air respectively at 101kPa absolute pressure. Then, the purge gas was switched into anode via 4-way valve for 20 seconds and switched out to vent for 20 seconds for 2 cycles. The gas purge residence time is defined as the volume of the anode flow channels divided by the purge gas flow rate. The gas purge residence time in these tests ranged from 0.08 to 1.35 seconds.

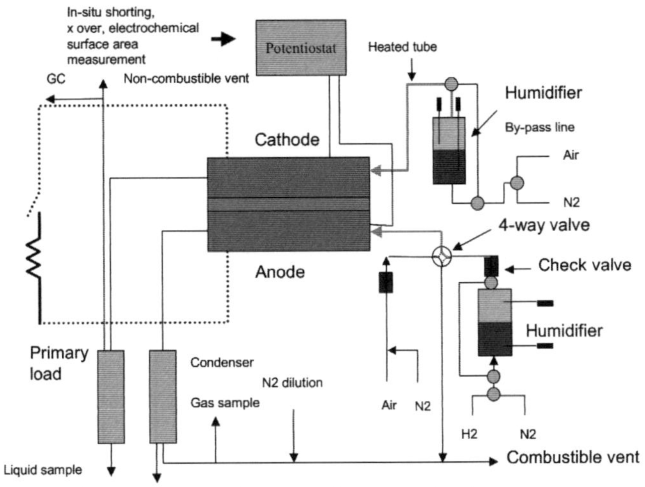

Figure 1. Schematic flow diagram of startup/shutdown test

A typical startup/shutdown test chart is shown in Figure 2. Zones 1 and 6 indicate cell voltage at 0.2 A/cm^2. Zone 2 is the cell at open circuit voltage. Zone 3 shows the cell voltage decrease as air started to purge through the anode, and zone 4 shows the cell voltage increase as H_2 started to refill the anode. There are 2 spikes in the zones 3 and 4 showing that 2 cycles were executed in each test. Zone 5 shows that the cell is conditioned by being potentiostatted at 0.6 V for 15 minutes before 0.2 A/cm^2 current is drawn.

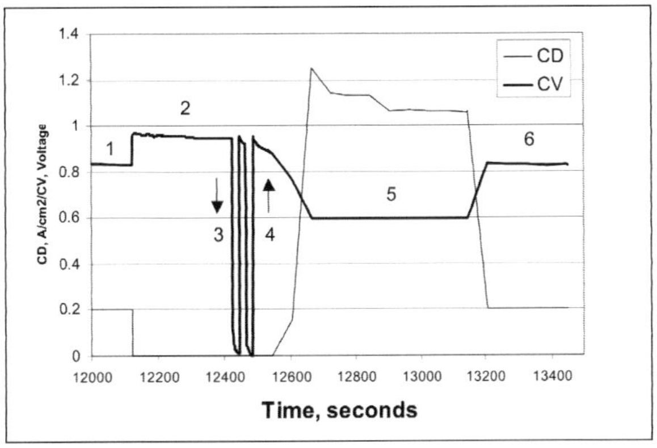

Figure 2. Current density (CD) and cell voltage (CV) changes during the start/stop test

Results and Discussion

Many series of start/stop tests in 50-cm^2 single cell were performed in order to understand the causes of cell degradation and to quantify the relationship between the cell degradation rate and the air purge residence time. Figure 3 compares the effects of air purge and of N_2 purge on the degradation of 50-cm^2 single cells. The results show that air/ H_2 purge caused very significant cell degradation compared to N_2/ H_2 purge, although N_2/H_2 still showed slight cell degradation. Under these conditions, more than 90% of the cell degradation observed was due to air/ H_2 fronts at the anode rather than to other effects which were also present in the N_2/H_2 experiments.

Figure 4, shows scanning electron micrographs of the fresh and post-test MEA's. In picture A (fresh), the thicknesses of the cathode and anode catalyst layers are quite similar (~12μm). However, in picture B, for which the MEA has been subjected to 200 start/stop cycles, the cathode catalyst layer shows a very significant thickness decrease by a factor of ~2 while the anode catalyst layer shows no significant changes. This evidence indicates that carbon in the cathode catalyst layer has corroded, according to:

$$C + 2\,H_2O \rightarrow CO_2 + 4\,H^+ + 4\,e^- \qquad E^o = 0.21\ VRHE \qquad [5]$$

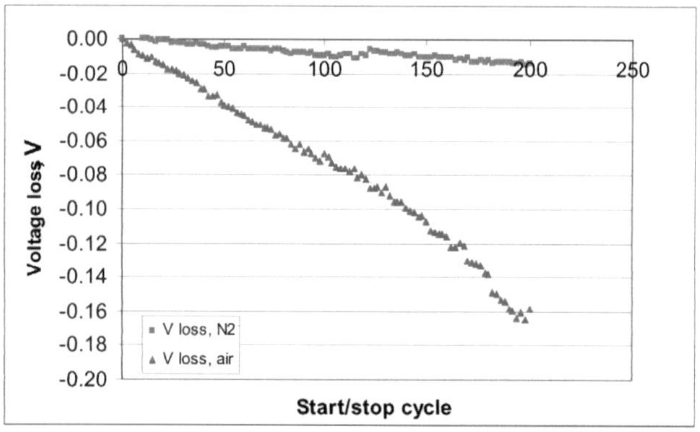

Figure 3. Comparisons of air/H_2 purge and N_2/H_2 purge on the degradation of 50-cm^2 single cell.

While this corrosion could thermodynamically occur even at the normal open circuit voltage (OCV), the absence of such corrosion in H_2/N_2 experiments, in which a similar OCV of ~1.0V RHE was achieved, points to a different explanation. In addition, potentiostatic experiments have shown only low carbon corrosion currents in acid below 1.2 V [6]. The major cause of the carbon corrosion in the shutdown/startup tests is believed due to a local high potential between parts of the cathode and the adjacent electrolyte while the air/ H_2 front or H_2/air front is passing through the anode when the cathode is still filled with air [2]. Figure 5 summarizes the proposed mechanism. The H_2/air front causes a short-circuited fuel cell to operate on the anode, resulting in a lateral flow of hydrogen ions from the H_2-flooded to the air-flooded portion of the anode. This high ion current through the high lateral ionic resistance of the membrane produces a significant (~0.5V) lateral potential drop in the membrane, leaving the portion of the cathode on the right side of Figure 5 poised at ~1.5V vs. the local electrolyte. This local high potential drives rapid carbon corrosion.

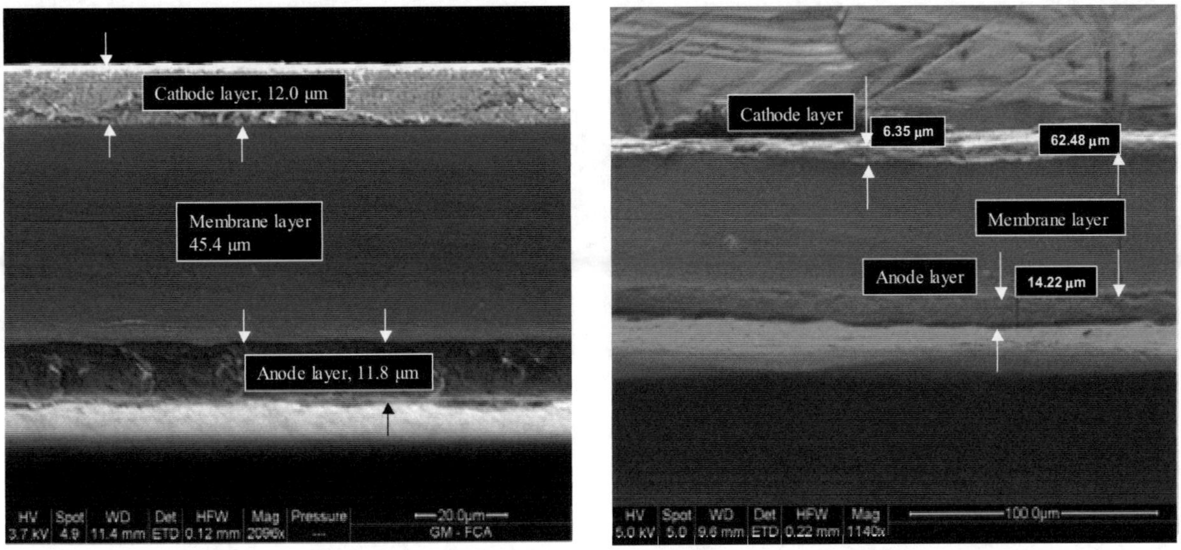

(A) (B)

Figure 4. SEM comparison of MEA sample before purge with air and after purge with air over 200 cycles at 80 °C. (A: Fresh sample, cathode layer on the top, B: Aged sample, cathode layer on the top.)

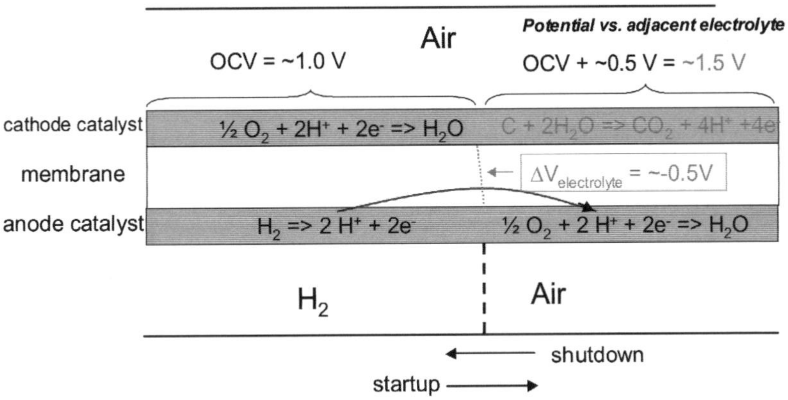

Figure 5. Mechanism yielding a local high potential on cathode while H_2/air front passes through the anode opposite an air-filled cathode

The effect of cycle number on the degradation of 50-cm^2 single cell is shown in Figure 6 for three different H_2/air front residence times per cycle. As was shown previously [2], the shorter the residence time, the lower is the degradation per cycle. Thus, degradation is related to the time that the air/ H_2 front remains in the anode compartment. By converting each cycle to the accumulated-front residence time, which is the front residence time multiplied by cycle number and by 2 fronts per cycle (start and stop), Figure 7 shows that short residence time and long residence time give a very similar trend of cell degradation when plotted against total accumulated-front time. This means that the carbon corrosion rate at this condition is similar no matter what the variations of the residence time are. The longer accumulated-front residence times allow more time for carbon corrosion. And, the more carbon corrosion on the cathode, the greater the cell degradation. In conclusion, the cell degradation controlling parameter is not the residence time per se; it is the accumulated-front residence time.

Plots of 50-cm^2 single cell degradation rates vs. residence time and 15-cell full size active area degradation rate (one point) vs. residence time are shown in Figure 8. Linear regression from the small-scale data shows that cell degradation rate per cycle is closely linear to the front residence time per cycle. The degradation rate of the short stack with a 1.35 second residence time is close to the extrapolation of the small-scale test data. This shows consistency between small-scale and short stack results.

These results suggest that within this residence time range, the cell degradation rate per cycle can be easily estimated. However, even with the residence time of 0.05 second (shortest possible using an air flow rate equal to nominal full-power air flow), the degradation rate is still about 0.1 mV per start/stop cycle. Taking into account the vehicular 30,000 start-stop cycle requirements, the results suggest that using air to purge out H_2 in the anode during shutdown and using H_2 to purge out air in the anode during startup could not, by itself, meet vehicular requirements. More corrosion resistant catalyst support (e.g., graphitized carbons [6] or other materials. [7]) as well as improved startup/shutdown procedures [2,3] are needed to adequately mitigate startup/shutdown losses.

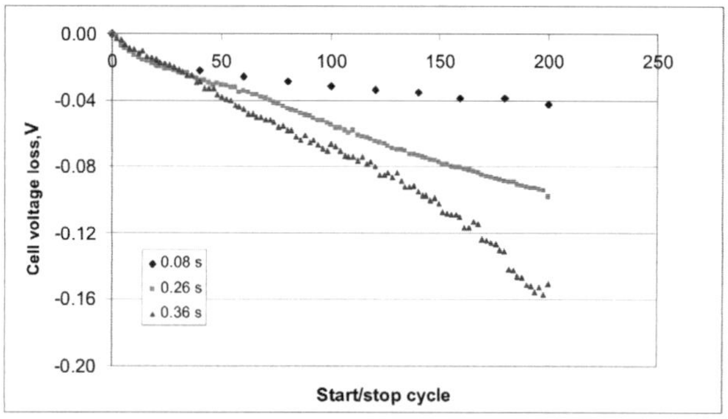

Figure 6. The 50-cm^2 single cell voltage degradation at 80 °C vs. numbers of start/stop cycles as a function of residence time

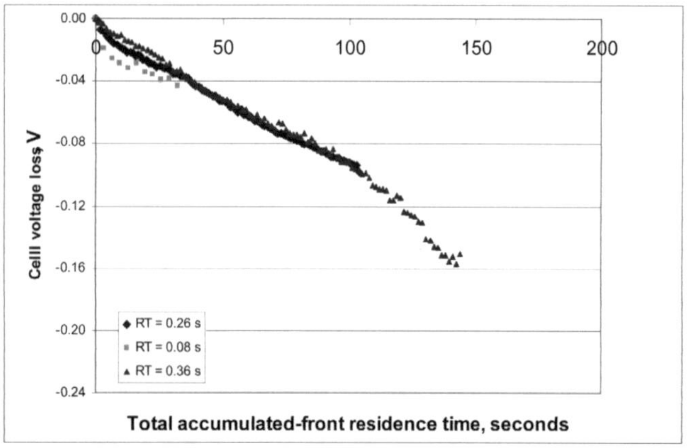

Figure 7. The effect of air purge residence time on the cell voltage loss as a function of total accumulated-front residence time at 80 °C

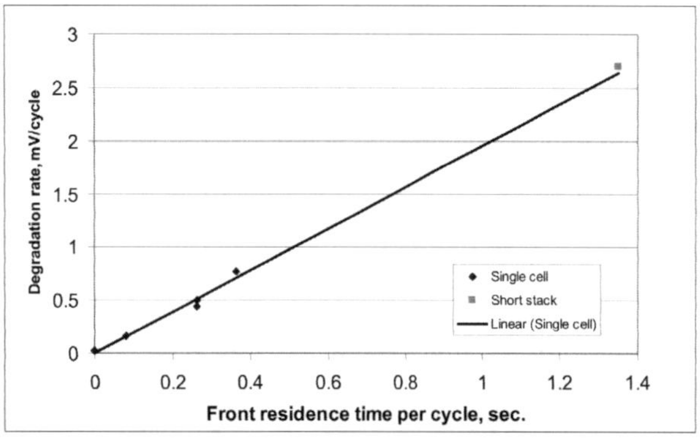

Figure 8. The 50-cm^2 single cell voltage loss as a function of front residence time per cycle at 80 °C

Summary

The effects of anode air purges on the degradation of single cells and short stacks at 80 °C were investigated in this paper. The observations include:

- Air/H_2 fronts at the anode while the cathode is filled with air result in significant decay of cell performance. This cell degradation was shown to be due to carbon corrosion at the cathode.
- The shorter the air purge residence time, the lower the cell degradation rate.
- The cell voltage decays are proportional to the total accumulated-front residence time and are consistent between the single-cell and short-stack results.
- To reduce startup/shutdown related voltage loss in automotive application, the development of stack startup/shutdown operation strategies and/or more corrosion resistance catalyst support materials are required.

References

1. Wilkinson, D. P. and St-Pierre, J., in: Handbook of Fuel Cells – Fundamentals, Technology and Applications, Edited by Wolf Vielstich, Hubert Gasteiger, and Arnold Lamm. Volume 3: Fuel Cell Technology and Applications, Chapter 47, page 611-626, John Wiley & Sons, Ltd., 2003.
2. Reiser, C. A., Yang, D., and Sawyer, R. H., Patent Application Publication, Pub. No.: US 2002/0076583 A1, 2002.
3. Yang, Deliang, Steinbugler, M. M., Sawyer, R. D., Van Dine, L. L., and Reiser C. A., Patent Application Publication, Pub. No.: US 2002/0102443 A1, 2002.
4. Landsman, D. A. and Luczak, F. J., in: Handbook of Fuel Cells – Fundamentals, Technology and Applications, Edited by Wolf Vielstich, Hubert Gasteiger, and Arnold Lamm. Volume 3: Fuel Cell Technology and Applications, Chapter 60, page 811-831, John Wiley & Sons, Ltd., 2003.
5. Roen, L. M., Paik, C. H. and Jarvi, T. D., Electrochemical and Solid-State Letters, 7 (1) A19-22, 2004.
6. Kinoshita, Kim, Carbon Electrochemical and Physicochemical Properties, John Wiley & Sons, Ltd., 1998.
7. Wang, J., Swain, G.M., J. of The Electrochemical Society, 150 (1) E24-E32, 2003.

Enzymatic Bio-Fuel Cell: A Promising Power Source

by N. B. Idupulapati and D. S. Mainardi, Louisiana Tech University, Ruston LA

Introduction

Biological Fuel cells, which convert biochemical energy to electrical energy, are proposed as alternatives to the current chemical fuel cells. An posibility is to develop enzymatic biofuel cells, which involve the application of oxido-reductase enzymes for oxidation and reduction of specific fuel at the electrode supports for the generation of electrical output.

We proposed to use Methanol Dehydrogenase as the anodic catalyst which is able to oxidize a wide range of primary alcohols as well as methanol. Methanol Dehydrogenase (MDH) is a bacterial periplasmic quinoprotein which has Pyrroloquinoline quinone (PQQ) [1]. The active site of this enzyme contains Ca^{2+} ion holding PQQ in position, which acts as a Lewis acid facilitating attack by the substrate on the PQQ [2].

In this work, Molecular Dynamics and Density Functional Theory simulations are mainly used to get insights into the dynamics of the active site in the presence of the fuel, and to obtain geometry, energetic, electronic configurations, and information regarding the dissociation of the fuel.

Procedure

Fully optimized geometry and vibrational frequencies of PQQ were obtained using Density Functional Theory (DFT) implemented by the program Gaussian 98, at the HF density functional theory level with 6-31G basis functions.

We have also used Molecular Dynamics (MD) simulations to get insights into the dynamics of PQQ and Ca^{2+} ion bonded to PQQ molecule. These simulations were carried out using the Merk MMFF94 force field at constant temperature of 300K, under a weak coupling heat bath system with constant temperature using the Berendsen thermostat. Once the thermodynamic equilibrium was reached, the simulation system approached the canonical ensemble (NVT), at constant number of particles N, volume V and temperature T, with periodic boundary conditions. The equations of motion were integrated using the Velocity Verlet method with a time step of 0.001 ps. The simulation length was 2000 ps with 600 ps equilibration steps and 1400 ps for the production stage.

Results and Discussion

Using the program Gaussian 98, at the HF density functional theory level with 6-31G basis functions we found that, the most stable molecular geometry, equilibrium bond lengths, vibrational frequencies, and Mülliken charges were obtained. The ground state spin multiplicity for PQQ is a singlet with a total energy equal to -1240.71715 Ha. Figures 1 illustrates the ground state configuration of PQQ and its IR vibrational spectrum. Charges of selected atoms are given in italic.

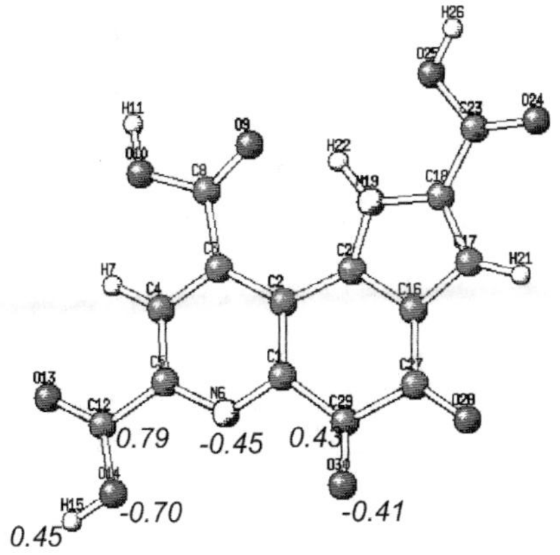

	Bond length(Å)
C5 – N6	1.3195
C5 – C12	1.4806
N6 – C1	1.3187
C12 – O14	1.3326
O14 – H15	0.9546

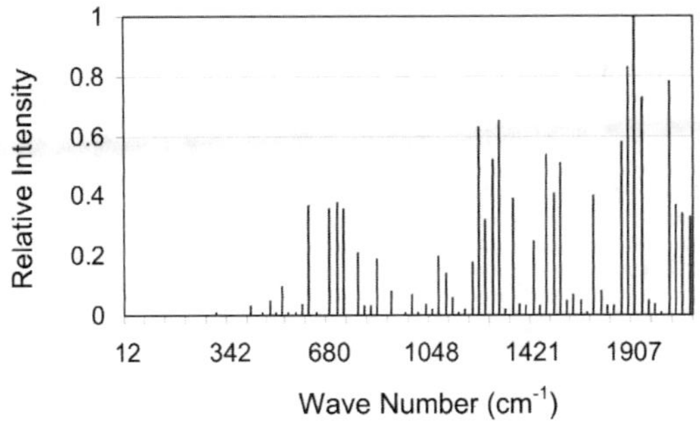

Figure 1. Ground state configuration of PQQ, selected bond length, and the IR vibrational spectrum at the HF density functional theory level with 6-31G basis functions.

Molecular Dynamics simulations of Ca^{2+} ion bonded to a PQQ molecule were carried out using the Merk MMFF94 force field at 300 K. Figure 2 shows the PQQ - Ca^{2+} ion interaction and selected bond lengths after 2000 ps.

It can be noticed that Ca^{2+} binds the PQQ molecule through the "cavity" formed by the negatively charged nitrogen and oxygen atoms, with the carbonyl group (C═O) facing up within the carboxyl (—CO_2H) group, as shown in Figure 1. However, when Ca^{2+} binds the PQQ molecule, the hydroxyl (—OH) and carbonyl groups within the carboxyl group switch psitions.

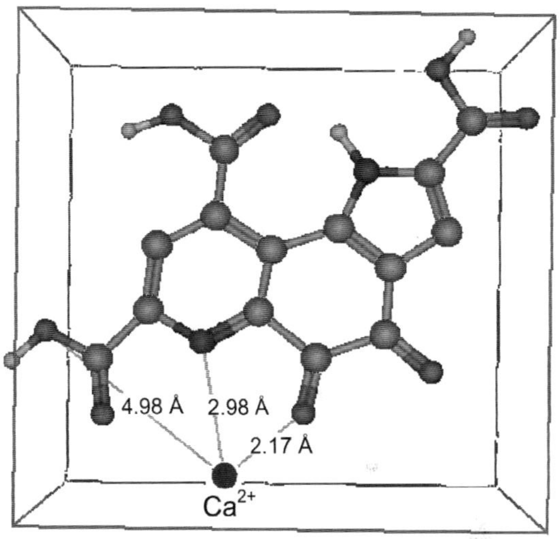

Figure 2. Molecular Dynamics simulation of Ca^{2+} ion binded to PQQ molecule at 300 K.

Conclusions

This study provides initial steps to get useful insights into the energetics and dynamics of the active site of Methanol Dehydrogenase, which can be used as the anodic catalyst of the bio fuel cell. The complete active site of this enzyme as well as the dissociation of methane will be considered in future work.

References

1. C.Anthony; Methanol Dehydrogenase, a PQQ containing Quinoprotein dehyrogenase, Sub-cellular biochemistry, Vol 35, 2000 , pages 73-118.
2. Christopher Anthony, Paul Williams; The structure and mechanism of Methanol dehydrogenase, Elsevier science, 2003. pages 18-23.

High Explosive Recrystallization Model and User Interface Development

Presented at the 2004 AICHE Spring Meeting
New Orleans, LA
April 25-29, 2004

Kenneth Schwartz, Tim Quinlin
BWXT Pantex

Evangelos K. Koutsavdis, Richard LaRoche, André Bakker
Fluent, Inc.
Lebanon, NH 03766

Background

Hexanitrostilbene (HNS) is recrystallized prior to use. The methods used for recrystallization were developed many years ago. The methodology for operations were based on operational limitations at that time. It is desired to have a tool to develop a scientific base for the run cycles of recrystallization that result in an improved purity and more desirable particle size. Using Fluent, a computational fluid dynamics (CFD) code, a model has been developed to allow the scientists to develop and optimize new and existing recrystallization operations while reducing the need for time-consuming and costly experimentation.

Methodology

As part of the High Explosive Modernization Effort at Pantex, computational models are being developed to enable a scientist to evaluate new methods for high explosive manufacturing. One such effort is the development of a computational model that can be used to evaluate HE recrystallization. This process was selected as an evaluation of the capability since recrystallization does not involve chemical reactions or the addition of other constituents, such as binders, and is thereby a simpler model to develop.

The intent of the model is to allow a scientist to perform the analysis without having to run experiments. Experimentation in a chemical reactor vessel results in large quantities of waste material. A desire of this model is to allow the scientist to use the model without expansive knowledge of the computational model. This resulted in 2 steps being performed to develop the model; develop a template to interface the scientist and the CFD code, and develop a model for recrystallization in the CFD code. The CFD code used for this effort was Fluent 6.0, and the corresponding mesh generation software, GAMBIT.

HNS Recrystallization

The HNS is dissolved in a solvent in a glass-lined Pfaudler stirred tank. The Pfaudler vessel is typically equipped with a 3-blade retreat curve impeller (RCI) operating at variable rotation speeds and a single baffle. The stirred tank is jacketed to enable heating and cooling of the contents. The stirred tank also may include various dip tubes and probes.

The stirred tank is operated in batch mode. The HNS and solvent are added to the vessel. A typical solvent is NMP. The contents are heated to completely dissolve the HNS in the solvent. Then, the stirred tank is cooled to induce crystallization of the HNS. The cooling cycle may be either a continuous cooling process or a cyclic cooling process. No chemical reactions occur in the batch stirred tank. No gases are introduced or produced in the process.

Template Development

Geometry and mesh generation can be very labor-intensive. In addition, increasing experience with CFD modeling is required to generate meshes and model. To alleviate the burden of requiring a scientist to become familiar with the CFD code and lessen the time expenditure of generating the model, a template was developed to allow the scientist to select the operating parameters, and automate the computational methods. The template is a customized graphical user interface that automates geometry creation, mesh generation, and CFD solution procedures. The template allows ease of use as well as the flexibility to run new models without long model set-up times. As much of the synthesis and recrystallization operations are performed in Pfaudler reactors, the first generation of this template is designed around their use. The most common options are preprogrammed into the template, however, the template allows for customized inputs from the user. After the geometry and operating parameters are selected by the user, the template will verify that the options do not conflict (e.g., impeller and baffle will collide).

The template steers the GAMBIT preprocessor to create the correct geometry and mesh. The FLUENT6 cases are three-dimensional (3D) and model the impeller-baffle interaction explicitly, either using sliding-mesh or multiple reference frame (MRF) methods. Therefore, no experimental velocity data is needed for the impeller discharge.

Single-phase flow field calculations using the MRF model only will take a few hours for a full 3D grid of approximately 500,000 cells. Transient heating or cooling calculations are more time intensive. The whole process of setting up FLUENT6 case files, performing the flow field calculation, and the temperature field calculation are part of the template.

There are two main options regarding the tank geometry; the user can choose from a predefined family of tanks with the RCI impeller or can create its own geometry. The built-in family of tanks includes six different volume tanks (30, 50, 100, 200, 500, 1000 gallons capacity). For each of these tanks, the user can select between an H-type of Fin-type baffle.

A large number of options exist for a custom tank. The user can select between a single and a double impeller system. The impeller options for the top impeller in a double impeller system or the blade in a single impeller system are: three and four blade pitched blade turbines, three and four blade hydrofoils, curved blade turbine, retreat blade turbines with standard blades and wide blades, or a vertical blade impeller. In a double impeller system, a curved blade turbine or a retreat blade turbine with standard or wide blades can be used for the bottom impeller. Depending on the location of the impeller, the following options exist for the baffle; fin baffle, D-style baffle, H-style baffle, or a finger baffle. A dip tube is optional.

A sample screen for the template is shown in Figure 1.

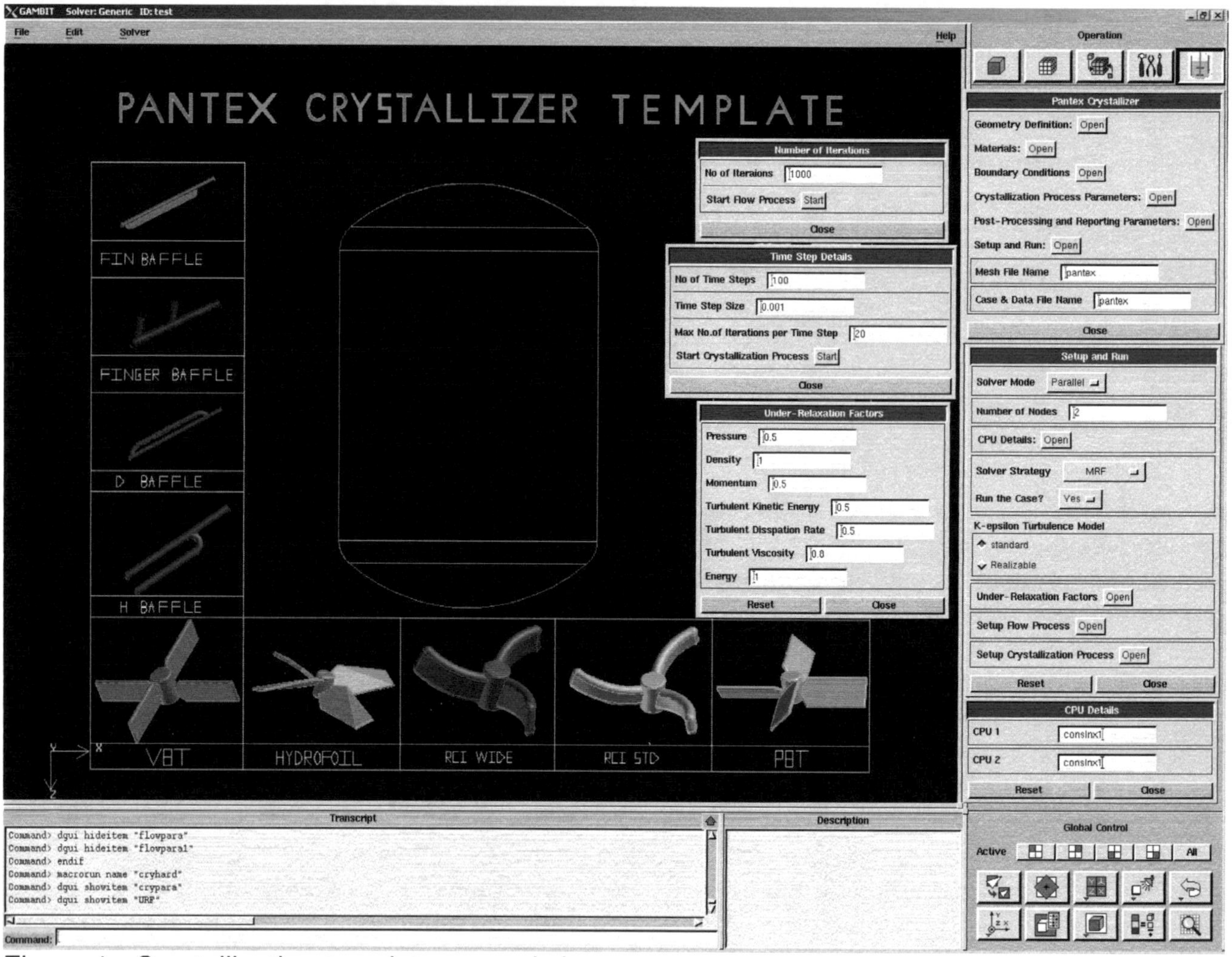

Figure 1: Crystallization template general view

Figures 2 and 3 show typical mixing tank calculations. Although the following flowfield depictions are not part of the capabilities of the template, they can be created as part of advanced post-processing based on the template generated FLUENT files.

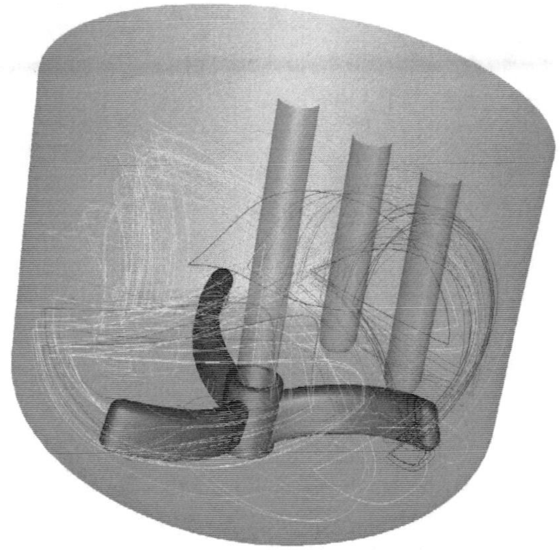

Figure 2: Predicted path lines in the tank.

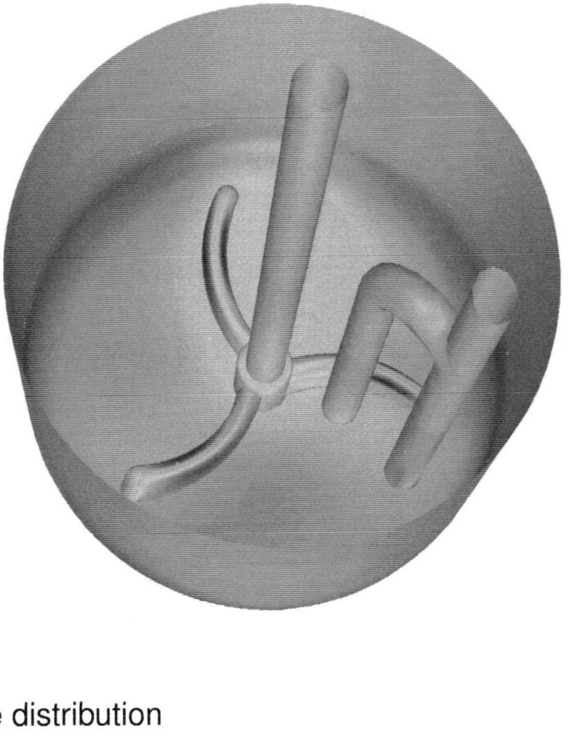

Figure 3: Predicted pressure distribution

Crystallization modeling

The driving force for crystallization is supersaturation. In this system, supersaturation is achieved by cooling the HE-solvent mixture, which decreases the HE solubility. From the fluid dynamics point of view, a rigorous approach would be to model the solid/liquid interactions with a multiphase model. The multiphase model approach is computationally expensive and outside Pantex's desire to have a quicker and easy-to-use engineering design tool. Therefore, we treated the crystal/solvent mixing as a single slurry phase.

Fluid flow modeling is already built into Fluent CFD. The crystallization models are developed through user defined functions written in C programming language. Because this model is to be of value to a scientist, it was decided to use the Standard Method of Moments (SMM). The value of this method is that the computation time will be reduced. The drawback is that the method is fairly new and the quantity of information that can be gained is not as specific. As work continues on this method, more information will be able to be obtained. At present, this model will be used to compare runs with one another, instead of as a tool to predict the output.

Standard Method of Moments

It is necessary to solve particle balance equations in order to determine crystal size distribution (CSD). The direct solution of the partial differential equations for the population balance is very expensive computationally. Therefore, the standard engineering approach is to transform the population balance using the standard method of moments (SMM) [1] (Wei, et.al 2001). The SMM offers the ability to calculate the mass averaged crystal mean size, d_{32}, the mass averaged total crystal mass, M_t and the mass averaged crystal variation coefficient, COV (standard deviation/mean). By coupling the SMM approach with CFD, the effects of local temperature and mixing on CSD can be taken into account. The transport equations for the 0th, 1st, 2nd and 3rd moments are solved as scalar transport equations with FLUENT6 using custom UDFs.

A moment is a statistical property of a distribution. The internal coordinate is a particle dimension. Since HNS is a needle-like particle, the length has been used.

$$m_j = \int_0^{+\infty} n(L;x,t)L^j dL$$

The standard method of moments can be used to determine moments of the number density function for those properties that we are looking for. In this case, we incorporate growth and nucleation terms into the method to evaluate crystal properties over time.

$$\frac{\partial m_j}{\partial t} + \nabla(um_j) = \nabla(\Gamma \nabla m_j) + O^j J + jGm_{j-1}$$ where

the growth term, G, is given by

$$G = k_G(c - c^*(T))^G$$

and the nucleation term, J, is given by

$$J = k_j(c - c^*(T))^J$$

In the above relation, C is the concentration of the solute and C^* is the saturation concentration. The concentration of the solute was determined by solving a convection diffusion equation for it:

$$\frac{\partial C}{\partial t} + \nabla(\vec{u}C) = \nabla(\Gamma_{j,eff}\nabla C) + S_C$$

where the sink term S_C is:

$$S_C = -\alpha_s \rho \rho_C M_2 G / MW$$

where α_S is the shape factor ($\pi/2$ for spherical particles),ρ and ρ_C are the density of the solvent and the crystals respectively, M_2 is the second moment, G is the growth rate and MW is the molecular weight of the solute.

The parameters Kg, Kj, g, J must be determined experimentally. The experimental determination of these parameters is being investigated at Pantex.

To solve for the properties we would like to determine, the SMM will evaluate the first four moments.

Mass averaged crystal mean size:

$$d_{32} = \frac{\sum m_{3,i} V_i}{\sum m_{2,i} V_i}$$

Mass averaged total crystal mass:

$$M_t = \frac{\pi}{6} \rho_c \frac{\sum m_{3,i} V_i}{\sum V_i}$$

Crystal Variation Coefficient

$$CV = \sqrt{\frac{\sum m_{2,i} V_i x \sum m_{0,i} V_i}{(\sum m_{1,i} V_i)^2} - 1}$$

Once a geometry is set up, the flow and energy equations for the reactor are solved at steady state. After convergence, the SMM are solved in an unsteady state.

Thermal Modeling

The computational modeling of crystallization involves the modeling of the unsteady heat transfer through the walls of the vessel during the cooling-heating cycle. Due to the complexities of unsteady thermal modeling and the additional computational cost that it involves, we assume that the solution is uniformly and instantaneously cooled or heated. The user supplies a file containing a table of temperature vs. time. Our model assumes that this temperature is the bulk temperature of the solution without any spatial variation due to convection or conduction through the walls. The effects of the complete thermal model will be considered for future investigation.

Testing

In order to test the above standard method of moments we wrote an independent code that implements a simplified SMM. This simplified SMM ignores the convection and diffusion terms in the SMM equations and thus should be applicable for mixing independent crystal growth. The simplified equations are:

$$\frac{dm_j}{dt} = 0^j \cdot J + jGm_{j-1} \qquad j = 0,1,2,3$$

and

$$\frac{dC}{dt} = S_C$$

The crystallization of NTO in water was used as a test case. Experimental data for the nucleation and growth kinetic orders and coefficients exist in ref [2]. For the template run we used a rotational speed of 500rm for a 30 gal vessel. The water quantity used was 20 gals. Ref [2] indicates that at 500rpm the crystal growth is mixing independent.

The crystallization kinetic orders and constants used are summarized in the following table:

Table 1: Growth and nucleation parameters and coefficients.

k_n(No./m^{3}sec)	n	k_g(m/s)	g
4.59E14	1.740947	0.00044	1.5334

Figure 2 depicts a comparison of the mean crystal size as predicted with the template model in comparison with the simplified SMM ordinary differential equation (ODE) model. The solution was cooled at a constant rate of 0.7C/min from 60C to 40C. A time step of 1 sec was used for both calculations. The initial concentration was 0.43 Kmol/m^3. We see the very good agreement between the two models. The total computational time was 2 CPU days for the template and a few seconds for the ODE model.

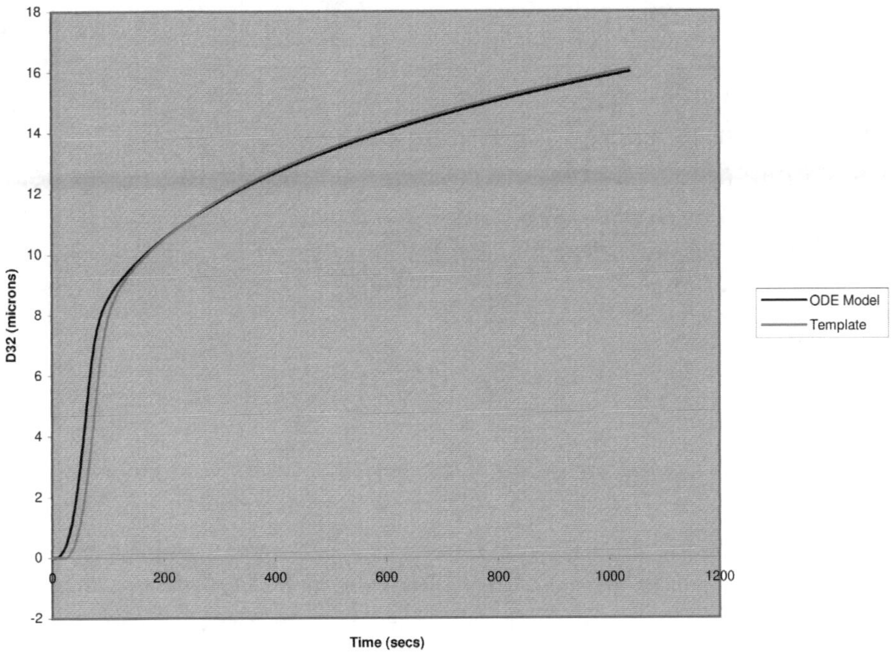

Figure 2: Comparison between the template model and the ODE model.

A comparison was made of the above results with experimental results for ref. [2]. The agreement was not favorable. This may be from the differentiating factors, such as the size and shape of the impeller and the vessel as well as the initial temperature upon which the kinetic orders and coefficients were based[3]. Validation of this model will have to be performed through experiments at Pantex.

Model Limitations

The following limitations should be considered for this model:

- Lack of comprehensive testing. At present, no data is available that would allow us to test the above model with vessels and processes used. Although the present model can give valuable information about the effect of different parameters on the crystallization process, it should not be considered as the primary tool for determining the final crystal distribution.
- Thermal modeling. More accurate thermal modeling is needed that would take into account convection and conduction through the vessel walls. Inclusion of thermal modeling is expected to increase the computational cost of a simulation.
- Discrete binning. The model was developed using the standard method of moments. As such, only a single distribution can be assumed for crystal sizes. Therefore, enough needs to be known about the recrystallization to know the importance of additional peaks in crystal size. Until the quadrature method is implemented, this tool should only be used as a comparative method.
- Computational cost. For crystallization processes that last days, the above modeling could be computationally expensive.

Future Work

Ongoing efforts include revisions and modifications to the template and recrystallization model. These efforts include:

- Testing of the current template with experimental data.
- Experimentation to determine physical constants required for the models
- Addition of HMX and RDX recrystallization models.
- Implementation of the quadrature method of moments.
- Modularize the template to allow inclusion of various models as they are constructed.

References

[1] Wei, H., Zhou, W. and Garside, J., 2001, CFD Modelling of Precipitation Process in a Semi-Batch Crystallizer, *submitted for special issue of IECR.*

[2] Kim, Y. H., Lee. K., Koo, k. k., Shul, Y.G, Haam S., "Comparison Study of Mixing Effects of Batch Cooling Crystallization of 3-Nitro-1,2,4-triazol-5-one (NTO) using Mechanical Stirrer and Ultrasound Irradiation", Cryst. Res. Technol. 37, 2002, pg. 928-944.

[3] Kim, Y. H., Private communication, 9/24/2003

Application of CFD for Improved Vertical Column Induced Gas Flotation (IGF) System Development

Chang-Ming Lee and Ted Frankiewicz
NATCO Group, Inc., 2950 North Loop West, Suite 750, Houston, TX 77092

Prepared for presentation at AIChE Spring National Meeting 2004

ABSTRACT

NATCO has completed the design and development of a single-cell vertical column flotation system for use in treating produced water on floating oil and gas production platforms. Computational Fluid Dynamics (CFD) simulations helped identify the cause of performance problems with early versions of column flotation and provided the insight required for design improvements. A combination of CFD simulation and physical model testing significantly reduced the time and cost required to develop the enhanced design. The physical testing validated the results of the first part of the CFD study and suggested additional modifications that were then included into later CFD simulations for further improvement. The development program culminated with the successful operation of a 10,000 BWPD column flotation system installed on a spar platform in the Gulf of Mexico.

VERTICAL COLUMN FLOTATION

Induced gas flotation (IGF) is a technology used to clean up oily water that is co-produced along with oil and gas. Oilfield exploration and production companies continue moving to ever-increasing water depths to produce oil and gas reserves and thus cause the shift from traditional fixed-leg platforms to floating structures such as the TLP (Tension-Leg Platform), SPAR and FPSO (Floating Production, Storage and Offloading). Due to space restriction on floating platforms and high cost of their construction, the oil and gas industry increasingly applies IGF in the configuration of a single-cell vertical column rather than the conventional four-cell horizontal design for onshore processing.

Flotation cells work by hydraulically introducing small gas bubbles, typically in the range of 100 to 500 microns, into the produced water stream. Column flotation units use a vertically oriented cylindrical vessel for the produced water treatment. Figure 1 shows the general layout of the vertical IGF Column. The produced water enters at the upper part of the vessel and flows downward. Gas bubbles are introduced into the lower portion of the vessel and float upward. The gas bubbles attach to and/or drag the oil and oil-coated solid contaminants and together they rise to the liquid surface on the top of the vessel. The agglomerated contaminants are removed by a skimming mechanism and treated clean produced water exits from the bottom of the vessel beneath the point of gas entry. The clean water can then be sent to overboard disposal.

NEW PRODUCT DEVELOPMENT

Instead of using a sparging tube, the formation and distribution of gas bubbles inside a vertical IGF Column can be better achieved by using a properly designed eductor that uses recycled, clean produced water as the motive liquid. Gas from the top of the vessel is routed to the eductor through a small diameter tube. As the clean produced water is pumped through the specially designed eductor, a low pressure zone is created in the eductor and gas is pulled into the eductor by the vacuum and mixed with the clean water prior to entering the vessel (basically a controlled Venturi effect). The gas/water mixture enters at the lower part of the vessel and encounters the oily produced water in a countercurrent flow pattern. Experience has shown that it is extremely difficult to obtain a uniform dispersion of gas bubbles in a column flotation unit for effective mixing with contaminated produced water, thus necessitating the CFD-assisted design for this purpose.

In order to reduce the cost and time required to develop an improved single-cell vertical column IGF technology, NATCO engineers applied FLUENT Version 6 with the Eulerian Multiphase Model running an unsteady state segregated solver. Flow of water and gas bubbles with various sizes were simulated within the IGF unit. Variables studied in this development project included the Gas Volume Fraction (GVF) introduced into the flotation cell through the eductor, the size of bubbles in the gas/water mixture flow through the eductor, the geometry of the device used to introduce contaminated water into the IGF, and the geometry of internal baffles, which assist optimized bubble distribution within the flotation cell, as well as various means to suppress wave induced sloshing near the gas-liquid surface in the skimming section of the IGF vessel.

The initial CFD simulations showed that the use of a traditional header and lateral inlet device for produced water causes a fluid short circuiting path to develop. This problem was resolved by replacing the inlet distributor with a cyclonic inlet device. The produced water exits the cyclonic inlet device in a radial swirling pattern and the water is smoothly distributed across the IGF vessel. This avoids the development of a fast down-coming stream from dragging the oily contaminants along the vessel wall and directly toward the bottom water outlet while by-passing contact with the flotation gas bubbles. Through the use of particle tracking during the post-processing phase of the CFD analysis, it was possible to estimate the average size of oil droplet, solid particle or gas bubble that will rise in the vessel against the net downward flow of the produced water.

The following CFD simulations predicted, and later physical testing with a test vessel (60" I.D. x 14' Height) conducted at NATCO's Gloucester, UK, laboratory verified, that the bubble distribution obtained from earlier versions of the column flotation technology were exceedingly poor. As illustrated in Figure 2, using a commercially available eductor, the larger gas bubbles released from the eductor mostly remained within a fairly small cylindrical area and formed a straight fast-rising plume as they travel up to the top of the IGF vessel. This poor distribution of the gas bubbles across the vessel limits the contact of gas bubbles with contaminants in the produced water stream and adversely affects the performance of the flotation cell.

To improve the spreading of bubbles into a uniform pattern that would contact most of the cross section of the IGF, the volume of water flowing into the cell through the eductor, the

GVF in the eductor flow, and the geometry of the eductor were all varied in the subsequent CFD simulations. Based upon the CFD simulation results, an improved internally-mounted Radial Eductor™ design was developed that, coupled with properly designed internal baffles, provides a uniform distribution of properly sized gas bubbles throughout the entire IGF vessel (see Figure 3). The CFD simulations also identified the presence of recirculation zones within the IGF that were previously suspected but impossible to detect in physical experiments. Such recirculation zones severely hindered the physical effectiveness of the IGF unit. The impact of the recirculation zones were minimized based on the results of further CFD simulations. These insights were used to modify the hardware design for subsequent physical testing.

Since this IGF system is primarily designed for service on floating offshore production platforms, additional design improvements were developed by applying a UDF (user-defined function) in FLUENT to simulate the vessel motion and the consequent liquid sloshing induced by the six degrees-of-freedom ocean wave motion. This liquid sloshing, if uncontrolled, could affect the performance of the IGF unit during storm conditions. After carefully examining several different configurations, an improved wave motion suppression baffle system was installed in the skimming region inside the offshore version of the IGF system. In the first stage, oil and water flow together into a pre-skim chamber from which water can egress through bottom openings while floating oil is retained. This accumulated oil layer is then skimmed into a conventional oil bucket with a limited amount of water carry-over. The two-stage skimming system ensures efficient contaminant removal is possible during calm and stormy sea states as the IGF vessel moves along with the floating platform.

RESULTS

Table 1 illustrates actual field performance data from a 10,000 BBL/Day offshore IGF system installed on a floating platform in the Gulf of Mexico. As indicated in the table, the effluent of the IGF is well below the federal limit (29 ppm) for liquid discharges into the ocean. Good field performance has been achieved under most sea state conditions, including those when considerable fluid sloshing inside the unit would be expected. Successful operation since start-up in early 2003 has verified the capability of the new vertical column flotation design.

ACKNOWLEDGEMENTS

The authors wish to acknowledge the technical support from Dr. Jayanta Sanyal (Fluent Inc., NH) and Dr. Srinivasa Mohan (Fluent India) and the hard work and contribution from NATCO project team, Neville Lange, Kevin Juniel and Stafford Menard, who have made it possible to complete this study.

REFENENCES

K. A. Juniel, T. Frankiewicz, and C-M. Lee, *Novel Vertical Column Flotation Design for Gulf of Mexico Deepwater Floating Structures*, 14th Annual Produced Water Seminar, Nassau Bay, TX (January, 2004).

TABLE 1. Offshore VersaFlo™ Field Performance Data

Date	Flow Rate (BPD)	Wave Height	Inlet Oil Content (ppm)	**Outlet Oil Content (ppm)**
02-07-2003	7000	8'	52	**27**
02-07-2003	6500	8'	48	**19**
02-07-2003	7000	10	37	**13**
02-08-2003	5500	8'	35	**13**
10-04-2003	6800	5'	47	**15**
10-07-2003	7000	5'	55	**23**

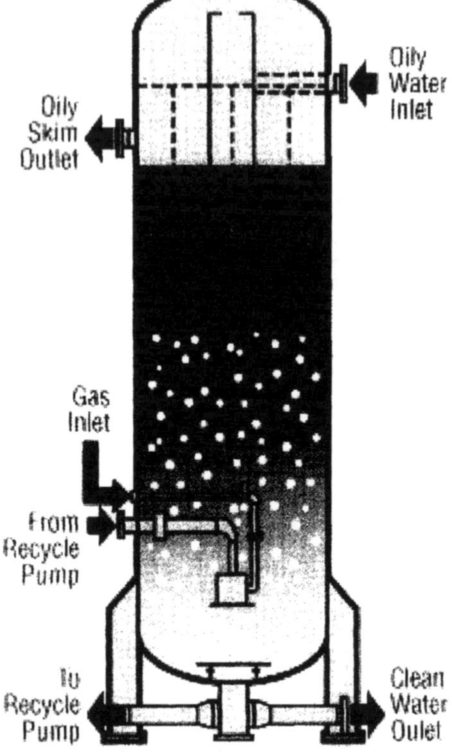

Figure 1. Schematic and Photo of a VersaFlo™ Vertical Column Flotation Unit.

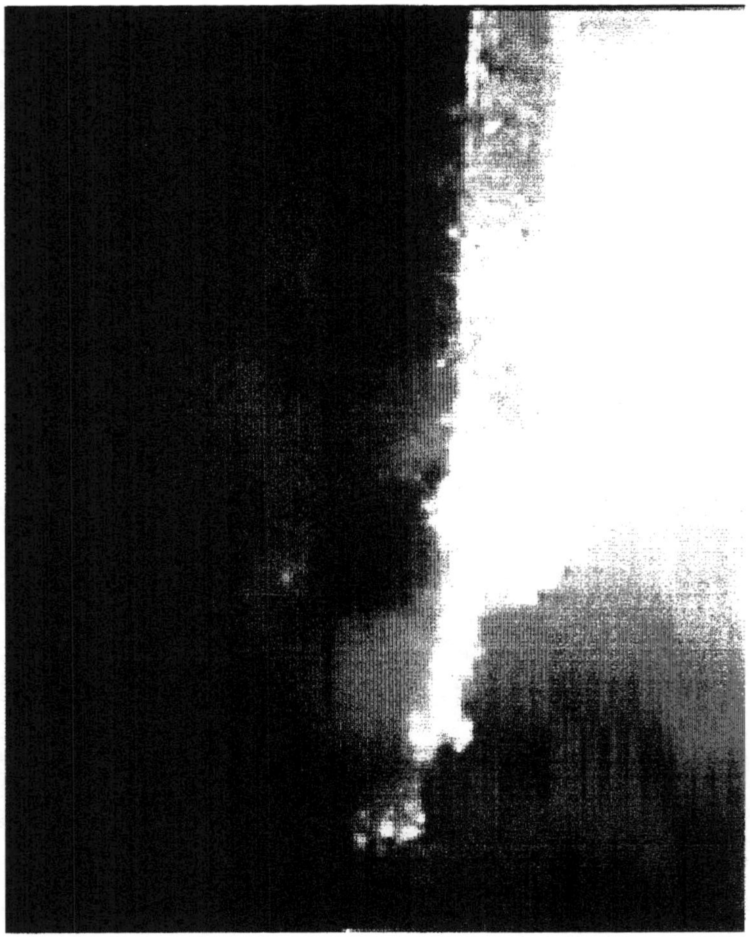

(a) Experiments conducted by laboratory at Gloucester, UK using *Megacell Eductor*

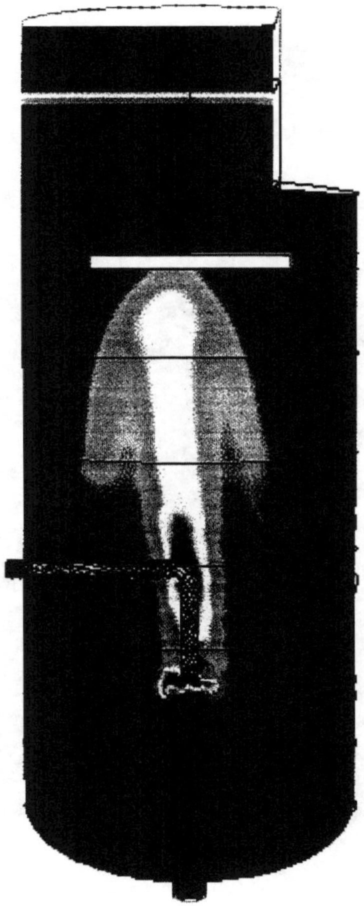

(b) CFD simulation result plot of contour of air volume fraction

Figure 2. Comparison of CFD simulation and physical laboratory testing of gas released by a commercially available eductor used in traditional gas flotation service.

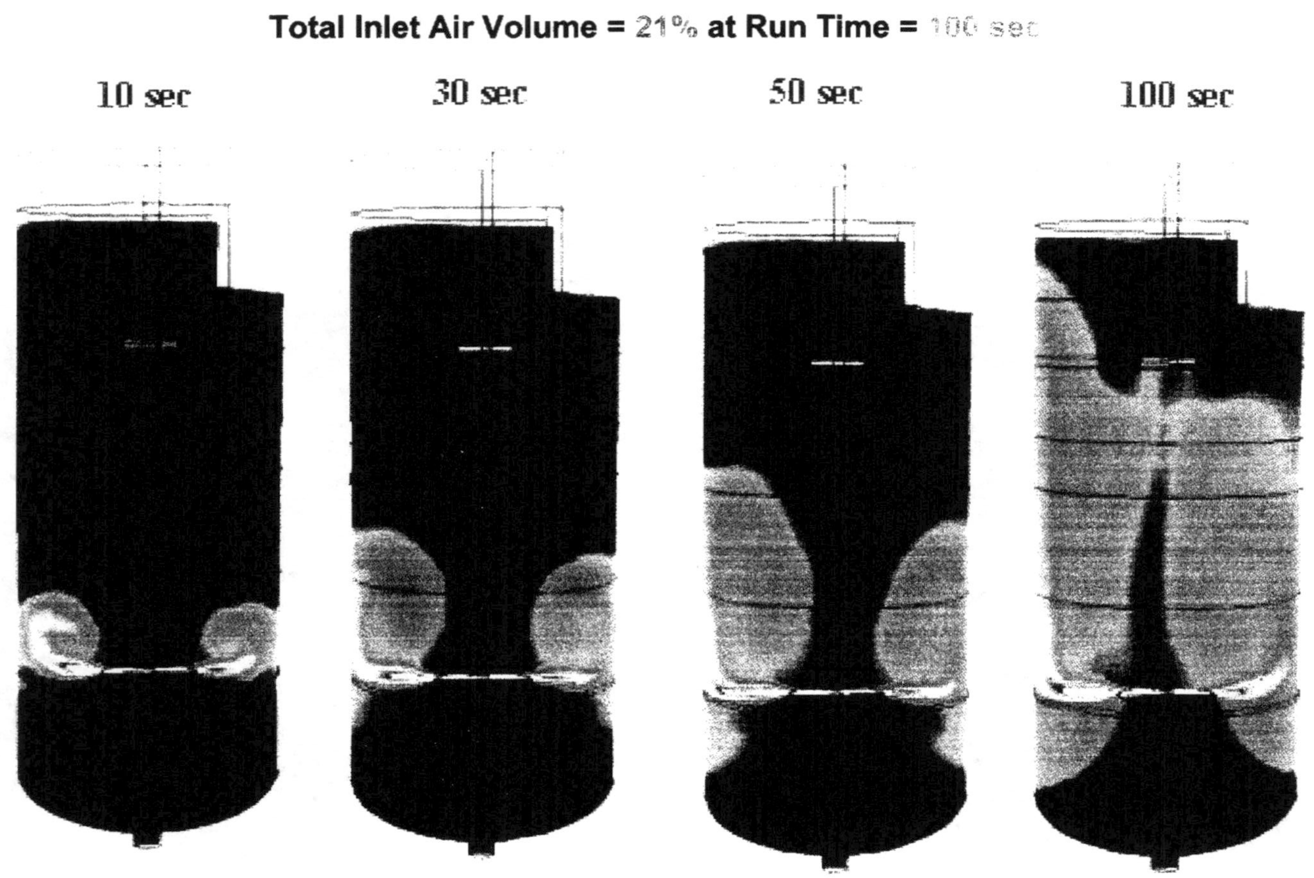

Figure 3. Improved gas bubble distribution provided by the (patent pending) radial eductor designed especially for column flotation service.

Superinsulation in Cryogenic Applications

James E. Fesmire
NASA Kennedy Space Center

Introduction

In today's world, the use of cryogenics and low-temperature refrigeration is taking a more and more significant role. From the food industry, transportation, energy, and medical applications to the Space Shuttle, cryogenic liquids must be stored, handled, and transferred from one point to another. To minimize heat leaks into storage tanks and transfer lines, high-performance materials are needed to provide high levels of thermal isolation. Complete knowledge of thermal insulation is a key part of enabling the development of efficient, low-maintenance cryogenic systems. The need for insulation is never a direct one. What is important is to save money on the energy bill or to be able to effectively control a system. Thermal insulation systems therefore provide energy conservation and allow system control for process systems.

Space launch and space exploration are both energy intensive endeavors. Cryogenics is called upon to meet this demand. Insulation is then needed to make the cryogenics effective. The term "superinsulation" has a number of different meanings to people in different technical areas. To the cryogenic engineer superinsulation typically means many layers of alternating reflective films and low-conductivity spacers, or multilayer insulation (MLI). Vacuum insulation panels for appliances are sometimes referred to as superinsulation. House construction using straw bales is also called superinsulation. A fitting analogy between house superinsulation and cryogenic superinsulation is given below:

> Superinsulation is more than simply piling up large amounts of insulation in the ceiling. It is an integrated system of building practices and components that achieves very low energy use by careful design, selection and installation of all of the elements that go into a home: walls, ceilings, vapor barriers, floors, windows, ventilation, and heating and cooling systems. All these building elements are carefully designed to make a superinsulated home virtually airtight, so that both heating in the winter and cooling in the summer are more efficient. Close attention to details and careful workmanship are required during construction to ensure that the installed insulation performs up to its full potential. [The Kentucky Energy Saving Home, Kentucky Division of Energy, 11/23/2003]

The point here is to emphasis the importance of a well-designed and properly executed system of thermal insulation. The extreme environmental conditions imposed by cryogenic systems lead to even greater and more complex technological problems that must be solved.

Materials

The overall efficiency of a cryogenic thermal insulation system can be summarized by the following four factors: 1) thermal conductivity, 2) vacuum level, 3) density or weight, and 4) cost of labor and materials. Materials typically come in three basic forms: bulk fill, foam, or multilayer. The vacuum level, or cold vacuum pressure (CVP), is the major cost driver for the design, fabrication, and maintenance of most systems. After the actual operating conditions are considered, an analysis of the total heat leak of the mechanical system is needed to determine the insulation requirements. Often only a common-sense thermal review of the system is needed to ascertain which level of insulation material should be selected. The performance level will dictate the insulation materials and mechanical support structures or joining devices to be used.

An insulation material's performance under a large temperature difference is given in terms of milliwatt per meter-kelvin (mW/m-K) and is referred to as the apparent thermal conductivity or k-value. To compare k-values for different materials one must understand the warm and cold boundary temperatures, the vacuum level, the residual gas composition, and the installed thickness. The designer has a very wide range of k-values with which to work: as low as 0.03 mW/m-K for perforated MLI blankets up to approximately 40 mW/m-K for cellular glass. As in all good designs, the performance must justify the cost. The performance of the total thermal insulation system as it is actually put to use is defined as the overall k-value for actual field installation or k_{oafi}.

Testing

Several test methods are usually needed to adequately test and evaluate the overall performance of an insulation system. Standardized material test methods can be employed for basic thermal, mechanical, and compatibility properties. Cryostat test methods provide the apparent thermal conductivity values for the insulation systems. Prototype testing is then needed to determine the actual performance for a specific mechanical system. The use of MLI systems illustrates the need for this three step testing process. The k-value for an MLI system under ideal laboratory conditions may be around 0.05 mW/m-K while the k_{oafi} can easily be 10 times worse.

The Cryogenics Testbed at Kennedy Space Center is active in research testing and support for a number of programs and initiatives for both NASA and commercial customers. Using a family of cryostat devices, a relatively quick and efficient way to characterize the thermal performance of materials under actual-use, cryogenic-vacuum conditions has been developed. The new test methods and equipment have been proven through nearly 1,000 tests of over different 100 material systems.

One target has been the development of insulation systems that perform well in soft vacuum, filling the wide performance gap between high vacuum systems and no vacuum systems. The soft vacuum systems represent a substantial new market area for consumer products and process applications by eliminating the burden of high vacuum fabrication and maintenance. A new Layered Composite Insulation (LCI) for double wall piping or vessels gives superior thermal performance at soft vacuum levels.

Applications

Applications of cryogenic insulation systems can be divided into three main categories according to the vacuum environments in which they operate. These three categories of cold vacuum pressure (CVP) are listed as follows: below 0.0001 torr or high vacuum (HV), from about 1 to 10 torr or soft vacuum (SV), and about 760 torr or no vacuum (NV). Materials used in high vacuum systems include, for example, MLI, micro-fiberglass, fine perlite, LCI, vacuum panels, and aerogels. Materials used in the newer soft vacuum systems include aerogels, LCI, and vacuum panels. Materials for no vacuum applications include foams, cellular glass, perlite, aerogels, and many others.

Examples of high vacuum systems include launch pad liquid hydrogen storage tanks, Shuttle fuel cell tanks, and thermal isolation for superconducting particle accelerators. Soft vacuum systems include propellant transfer piping connections, cost-efficient process piping, and storage of liquid oxygen on the surface of Mars. Examples of no vacuum systems include the Shuttle External Tank and liquid natural gas storage tanks. Each of these applications requires a combination of materials, mechanical structures, and equipment in order to function properly for the lifetime of the system.

A number of new materials are now commercially available for cryogenic thermal insulation application. These new materials include aerogel blankets by Aspen Aerogels (Pyrogel® and Spaceloft®), aerogel beads by Cabot (Nanogel®), and polyimide foams by Sordal (SOLREX®) and Inspec Foams (SOLIMIDE®). Other materials and composites under development are nearing the commercialization phase.

Technologies and markets forecast for rapid expansion into the 21st century, such as superconducting power distribution and hydrogen-based transportation, will require more efficient approaches to energy management for a wide variety of low-temperature applications. Cryogenic superinsulation technology is expected to be a foundational support to these developments.

Simplified Regenerative Heat Transfer Design

Louis Herrington, P.E.
LEHCO
112 Wildoak Dr.
Daphne, AL 36526

This paper describes simple and understandable equations for the design of Regenerative Heat Transfer Systems. Many bed designers rely on the complex time and temperature heat balance equations, Reference 1. Debora, et al, in Reference 1 used the energy balance equations to develop a time and temperature profile through the inert particulate bed that resulted from a flowing fluid of a different temperature. Because of the complexities involved with the equations and the subsequent lack of understanding of the results, the Plant Engineer is likely to neglect energy savings from hot gas streams. Other designers rely on Schumann's Curves, Reference 2, but they also fail to provide understanding and offer no clue as to accuracy. This paper describes a method that will not only replace Schumann's Curves but will provide basic bed design criterion that is quite comprehensible. With improved understanding of Regenerative Heat Transfer will come significant energy savings. From this analysis new light is given to Schumann's solution. All discussion and derivations are for the case of hot fluid and cold bed. Equations for cold fluid and hot bed will be included without comment.

Double S-Curves

The most important concept for the design of Regenerative Beds is a simple heat balance. Before proceeding with the heat balance it is important to reexamine the temperature Transition Region (TR). A heat balance will not appear quite useful until it becomes clear how the two concepts are related. After sufficient time, when hot gas passes through a colder bed or a cold gas through a hot bed, a region will exist between the hot and cold parts of the bed. That transition for both bed and fluid will take the form of a double S-Curve. Figure 1 illustrates and example of the two S-Curves of the TR. The TR will travel through the bed in the direction of the gas flow. Let the leading S-Curve be the hot fluid and the trailing S-Curve represent the cold bed. The beginning and ending of both S-curves will come together. It is easy to imagine this region to be long but the length of the region can be quite finite under suitable design conditions. All heat transfer to the bed occurs within the TR. Vertical lines connecting the two S-Curves represent the temperature difference between bed and fluid at any point within the TR. Area for heat transfer is provided by all particles within the TR. Slope of the TR is controlled by the initial bed temperature configuration.

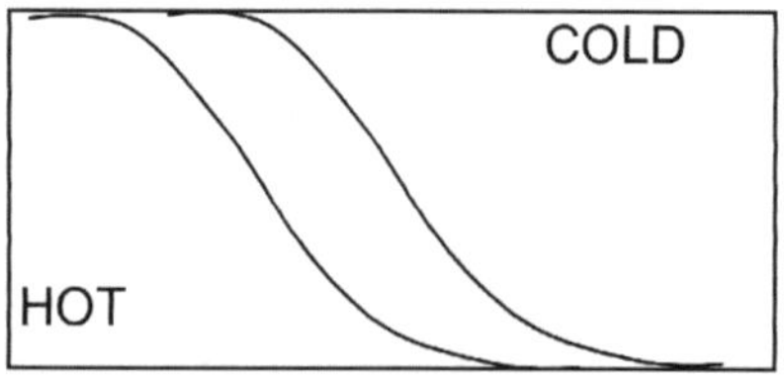

Figure 1: Transition Region

Transition Region Velocity

From a heat balance the velocity of the TR can be determined. A second element of the bed design is to estimate when hot gas of a selected temperature will exit the bed. In order to make that prediction the gas temperature profile is needed. As can be seen in Figure 1 that profile is complex. But there is a simple way to overcome that complexity. A relatively simple equation can be used to predict the initial gas temperature profile. A differential equation is derived and solved later that will predict the distance x along the bed for any gas temperature. This equation is easily derived for a uniform temperature bed because as the gas first moves through the bed it is always in contact with the cold bed.

The leading edge of the initial hot gas profile will provide an accurate estimate of the extra bed length needed for the TR. As an example to follow in the thought process, assign a quite reasonable TR velocity of 1 ft per hour for a selected gas flow. Generally the bed designer calls for bed heating to end as a selected fluid temperature exits the bed. It can be shown that the initial gas cooling temperature profile is established within seconds thus allowing the assumption that the profile exists at the start of heating. If the initial hot gas temperature profile indicates the selected temperature is 0.5 ft ahead of the inlet at the start of heating then add that increment of distance to the bed. If a 4 hour cycle is selected the required bed length becomes 4.5 ft.

Equations for initial uniform temperature bed design are listed below.

TR velocity in consistent units:

Equation 1 $v = Gcg/(bcs+\rho cg)$

Or Equation 1A, $v = GHg/(bHs+\rho Hg)$

Where:		
	v	TR velocity (Not superficial gas velocity)
	cg	Gas heat capacity
	cs	solid heat capacity
	$b=\sigma(1-e)$	bed bulk density
	ρ	Gas density at Tgo
	Hg	Gas enthalpy, Hhot-Hcold
	Hs	Solid enthalpy, Hhot-Hcold
	σ	Particle density
	e	Bed porosity
	L	Length of bed heated

Initial hot gas temperature profile:

Equation 2, Cold Bed $x = -(Gcg)/[h'(1-e)]Ln[(Tg-Tso)/(Tgo-Tso)]$

Hot Bed $x = -(Gcg)/[h'(1-e)]Ln[(Tso-Tg)/(Tso-Tso)]$

Where: (English units)		
	x	Distance along bed, ft
‘	h’	Gas/bed film coefficient, btu/hr-**ft3**-‘F
‘	Tgo	Initial gas temperature, ‘F
‘	Tg	Gas temperature at any point, ‘F

	Tso	Initial bed temperature, 'F
For later discussion	h	Gas /bed film coefficient, btu/hr-**ft2**-'F
	Ln	Natural logarithm

And h is related to h' by Equation 3.

Equation 3 $h'=h(Ap/Vp)$

Ap	Particle area, ft2
Vp	Particle volume, ft3

Therefore total bed length is given by Equation 4

Equation 4 $L = v\theta + x(\text{at selected exit Tg})$

θ Total heating time, hr

Calculation of the gas temperature profile verses x requires a value of h' as does the Schumann Curves. Equation 5 was recommended by Lof & Hawley and was used by Kern in Reference 3. Although it has questionable precision, it provides a source for values of h'. Since the major portion of the heated bed length can be independent of h' the accuracy of the coefficient is of reduced significance.

Equation 5 $h' = 0.79(G/D)$

Where: D Characteristic particle diameter, ft

Equation 5 is a dimensional equation and requires G in lb/hr-ft2 and D in ft. The equivalent spherical diameter was used in development of the correlation. Lof and Hawley developed Equation 5 from experimental studies with granite type gravel. Temperatures were relatively low and were mostly with air at about 200'F and ambient beds. After a review of other studies Lof and Hawley recommended a correction be made to the resulting prediction for higher temperatures. The recommended correction was the ratio of the central range temperatures raised to the 0.3 power.

If the Designer wishes to construct a TR profile for the bed additional equations are required. The TR will not be established until the front face of the bed is heated to near the hot gas temperature, Tgo.

Bed face heat up time is expressed by the easily derived Equation 6:

Equation 6 Cold Bed $t = -(cs\sigma/h')Ln[(Tgo-Ts)/(Tgo-Tso)]$

Hot Bed $t = -(cs\sigma/h')Ln[(Tso-Ts)/(Tso-Tgo)]$

Important Considerations

Regenerative bed designs of good accuracy will be available through the application of Equations 1, through 5. There will be opportunities to improve the results through better measurements and correlations of Heat Transfer Film Coefficients. Balakrishnan and Pei examined existing data in Reference 4 and concluded that no general correlation was available. For those who wish to review more recent efforts at predicting h a good start is the work of Thompson and Jacobs in Reference 5. Lof & Hawley in Reference 6 abandoned the plan they had developed and followed Furnes in Reference 7 by using Schumann Curve profiles to compare exit temperature profiles measured in experiments. In light of Equation 2 the method does result in approximate values of h'. One opportunity to compare methods was afforded because actual data from Run 10 of the study was included in Lof and Hawley's report. After examining the data it was apparent that the exit temperature profile was extended significantly due to heat losses to the surroundings. In spite of huge heat losses the exit profile was used as an overlay on the Schumann Curves for selecting the parameter called Y. When Equation 2 was applied to the data from Run 10 a higher value of h' resulted. That was not a surprise due to the effects of heat losses on Lof & Hawley's profile. Equation 5 resulted from Lof and Hawley's work and improvements are needed in accuracy.

In spite of using the volumetric heat transfer coefficient, the simplified equation derived in this paper to predict bed face heating in theory requires particles of near equal size. If not, the large particles would heat slower than the small particles and a range of temperatures would exist at the surface. There are compensating circumstances. A typical heating time for the face is somewhat long. Thirty minutes could be a typical heating time and some radial heat transfer will occur helping to bring the temperatures together. The simplifying assumption is Case 5 in McAdams, Reference 12, with zero conduction parallel to flow and finite radially.

Volumetric Film Coefficients

If a true coefficient of heat transfer is to be measured and correlated it should be adjusted from volume to area. Lof & Hawley, working toward a specific goal of characterizing gravel were not concerned about an area to volume ratio. That freedom will certainly not be available when particle shapes change. Present data shows that heat transfer resistance is predominately at the particle surface film. It is useful to review how the area to volume ratio for particles can vary. The simplest case is for spheres, with an area to volume ratio of 6/D. Cylinders with length equal to diameter are also 6/D. As cylinders lengthen, the constant begins to drop and approaches 4. Constants for flat plates can increase without bounds as thickness decreases. Hollow particles can impact the ratio in more than one way depending on the actual configuration. The K/D ratio effects heat transfer that passes from flowing fluid into particles, so factors such as stacking of flat plates will have a large effect. In some particles the K/D ratio for heat transfer will not likely be determined by other than experiment. Previous and future experiments with known K/D values such as spherical particles will and have been quite rewarding. Lof and Hawley presented experimental work by others. Other work by Sanders and Ford (Reference 8) with spherical particles found that h' correlated with G/D. Since h' is equal to hK/D the implication is that h is independent of D in the experimental regime. Differences in heat transfer characteristics for other types of particles might be best attributed to variations in K/D based on experimentally measured values of h'. A useful procedure for measuring h' will be illustrated later. Foust, et, al, in Reference 10 show experimental results from crushed ore samples. Surface area as a function of diameter is presented. Examination of the data showed K increasing with diameter. Correlations of data will require that such variables be considered. Particle properties other than specific heat have little correlation with experimental measurements of h because small particles achieve thermal equilibrium with surface temperature within seconds.

Application to Bed Design, Example 1

Application of Equation 1 - 5 can be made to a bed design. In Reference 3, Kern presented and example from Lof and Hawley, in which a bed of 1 inch gravel was heated by 60 lb/hr-ft2 of hot air. A bed length was to be determined that could absorb the heat from 200'F air that was available for 6 hours. Other needed properties were porosity, 0.45, gravel density, 165 lb/ft3, and bed heat capacity of 0.25 btu/lb-'F. Air density was listed as 0.0806 lb/ft3 and the air heat capacity given as 0.0191 btu/ft3-'F. In the Kern example, the solution was developed using the Schumann Curves that required the heat transfer coefficient h' in units of volumetric heat transfer. Equation 1 can be used to calculate the velocity of the TR. An examination of the development of Equation 1 shows that the air density required in the solution is for air at 200'F (0.0600lb/ft3).

TR velocity: v=60*0.237/(165*(1-0.45)*0.25+0.0600*.237)

Thus: v= 0.627 ft/hr

Then for 6 hours of heating the TR will move 0.627*6 or 3.76 ft. In the problem statement it was required that the hot gas emerge from the bed at 90'F after 6 hours. From Equation 2, the distance x from the bed entrance can be calculated for the 90'F point on the initial gas temperature profile. In order to make that application of Equation 2 requires the assumption that the leading end of the profile remains approximately intact. That is, the bed heats from the inlet and reshapes the upper temperature regions of the gas profile into an S-Curve while the leading end of the profile moves along with the TR velocity. While this will not be exactly true it offers a quite accurate estimate that is easily comprehended and well within design tolerance. Note the heated length calculation is rigorous. Saturated bed entry is stipulated for each cycle.

Equation 5 h' =0.79(60/(1/12))^0.7 = 79 btu/hr-ft3-'F

This value is substituted into:

Equation 2 x =- (60*0.237)/[79*(1-.45)]Ln[(90-50)/(200-50)]

x = 0.43 ft

Bed length via Equation 4 L = 6*0.627+0.43 = 4.2 ft verses 4.4 via Schumann.

There will be further comparison with Schumann's Curves later.

Derivation of Equation 1

From a simple heat balance the velocity of the TR, can be determined. To derive the expression for the velocity, consider a cold particulate bed of heat capacity cs, temperature Tso, and bulk density b. A hot gas of heat capacity cg, mass rate per unit area G, and temperature Tgo flows into the bed. At some point in the bed the hot gas has cooled to the temperature of the bed, Tso. Cooling is the result of heat transferred to the bed. After a time, in a perfectly insulated bed, the temperature of some portion of the bed is near the inlet gas temperature, Tgo. The region between where the bed and fluid are hot and cold is the TR. Now consider a heat balance across the inlet to the TR. A simplification of the heat balance allows the TR to become stationary. Then the heat enters the region from the left in the non-condensable hot gas and flows back to the left in a stream of particles. Because the TR actually moves in the direction of the gas the concept of a non-moving TR requires a reduction in the fluid velocity relative to the region. The reduction in velocity for the fluid is equal to the actual velocity of the TR. The velocity of the solid bed is then determined from the heat balance by reducing the heat input from the fluid as if it had slowed by the velocity of the solid. Heat in the fluid is Gcg(Tgo-Tso) less the reduced velocity component vρcg(Tgo-Tso), where the symbol ρ represents hot gas density. Heat leaving the region back to the left in the solid particles is equal to vbcs(Tgo-Tso). Equation A shows the heat balance.

Equation A $$Gcg(Tgo-Tso)-v\rho cg(Tgo-Tso)=vbcs(Tgo-Tso)$$

After canceling (Tgo-Tso) solving for velocity, v, reveals the velocity of the TR.

Equation 1 $$v=Gcg/(bcs-\rho cg)$$

The presence of ρcg in the denominator is somewhat mysterious because v can be derived as the heat, Gcg(Tgo-Tso) needed to saturate one foot of bed, bcs(Tgo-Tso). If ρcg is important to the equation it remains as an insignificant quantity in atmospheric gas problems.

As in all gas flow in beds, conditions should be selected to reduce the channeling of gases. Pressure drop in the bed is also a necessary consideration. Heat losses and heat capacity along the wall should be kept as low as possible for energy savings. As temperatures are dropping along the insulated bed resulting from heat loss to the walls there will be only small impact on the velocity of the TR. Any change will be expressed by Equation 1 or 1A where properties will be at the new temperature.

Derivation of Equation 2

To develop the initial gas cooling curve consider the small leading end of the hot gas as it flows through the cold bed. Let the small portion of gas have the mass dw. As it passes through the cold bed it cools by transferring heat to the cold bed. The incremental heat lost from the gas is dwcg(dTg) where cg is the gas heat capacity, and dTg is the differential change is gas temperature. The increment of heat transferred to the constant temperature particles in the increment of time dt is $h\bar{A}(Tg-Tso)dt$. In this expression $\bar{A}$ is the area of particles in contact with the thin leading edge dw of hot gas. Here the film coefficient is in units of ft2. The volume of the end increment is dV=dw/ρ=dwRTg/(PM) when gas density is expressed in terms of the ideal gas law and where M is gas molecular weight and R and P are the gas constant and pressure. For a unit cross-section of bed the particulate area $\bar{A}$ exposed to the gas leading edge dw is equal to ApNdV. Here Ap is the area of the average particle. The parameter N is the number of particles per unit volume. In more familiar terms N=(1-e)/Vp where e is the porosity and Vp is the volume of the average particle. Combining these expressions plus Equation 3, clearing terms, and rearranging gives the differential expression for temperature change with time for the leading edge of gas.

Equation B $$dTg/(Tg(Tg-Tso))=(h'/cg)(R/MP)(1-e)dt$$

Because of slow bed heating by the hot gas this profile is also the initial gas temperature profile. With the use of Equation B and numerical integration the value of Tg verses time, t, can be developed. Also, note the dependence of the derived equation on the beds uniform temperature. Equation B can be integrated using partial fractions. The integrated expression for time and temperature is shown in Equation C.

Equation C $(1/Tso)Ln[(Tgo/Tg)(Tg-Tso)/(Tgo-Tso)]= (-(h'/cg)(R/MP)(1-e))t$

The reader can verify from this equation just how quickly the temperature profile for the gas is formed. It required seven seconds to form in Example 2 to follow.

Subsequently, if use is made of the fact that $x=\upsilon dt$ in Equation B, where υ is superficial gas velocity, and Equations 7 and 8 are used to replace υ,

Equation 7 $\upsilon=G/\rho$

Equation 8 $\rho=PM/RT$, gas density

Where
υ Superficial Gas velocity
P System pressure
M Molecular weight of gas
R Universal Gas Constant

Then integration and substitution of Equation 3, $h'=h(Ap/Vp)$, results in a rearranged version of Equation 2

Equation 2(Rearranged) $h'/(cgG)(1-e)x=Ln[(Tg-Tso)/(Tgo-Tso)]$

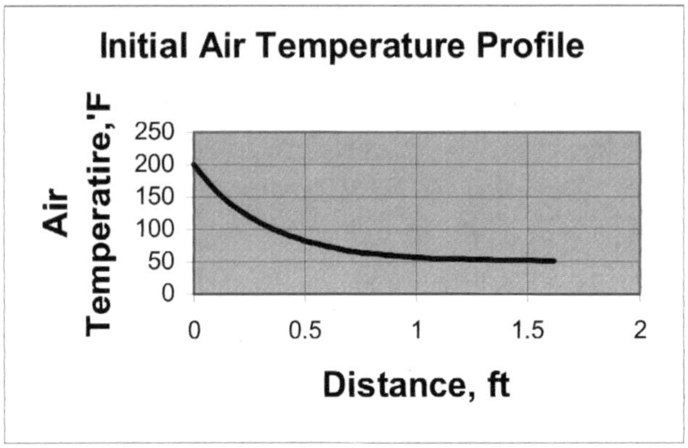

Figure 1. Developed from Lof & Hawley Data

Figure 1 illustrates the gas cooling profile for Example 1. After a short time Equation 2 is no longer correct. However, the leading part of the profile will remain closely intact and that is why it is useful for projection through the bed. Note that the Integral of $h'(1-e)(Tg-Tso)dx$ is equal to $Gcg(Tgo-Tso)$.

Example 2

Example 2 is also taken from the work of Lof and Hawley in Reference 5. Lof and Hawley made extensive measurements of heat transfer in a bed of length 3 ft, and 0.77 ft2 of area. Heated air was allowed to flow through the bed using several particle sizes and flows. Using exit temperature profiles and the curves of Schumann they reduced the data to obtain values of h' in units of btu/hr-**ft3**-'F, the coefficient of heat transfer between gas and bed. In one case the actual temperature profile was given. From that case referred to as Run 10, and Equations 1 to 5 it was possible to directly calculate a value for h' in units of btu/ft3-'F-hr. The data from Run 10 is presented in Table 1.

Time	(Tg-Tso)/(Tgo-Tso)	Tg, 'F
1	0.0109	67.46
1.5	0.058	73.77
1.75	0.14	84.76
2	0.262	101.11
2.25	0.397	119.20
2.5	0.555	140.37
2.75	0.687	158.06
3	0.79	171.86
3.25	0.861	181.37
3.5	0.908	187.67
3.75	0.946	192.76
4	0.967	195.58
4.24	0.984	197.86

Table 1. Data from Lof & Hawley, Ref.3

Other Bed Parameters

G	204.5 lb/hr-ft2
cg	0.2379 btu/lb-'F
σ	165 lb/ft3
cs	0.25 btu/lb-'F
e	0.454
Tgo	200 'F
Tso	66 'F

The volumetric film coefficient can be calculated by recognizing that the temperature point that emerged after 1 hour had moved along in the bed for all that time. Knowing the TR velocity allows the starting point in the bed for that temperature to be calculated. By substitution in Equations 1, 2 & 4,

$$v=Gcg/(\sigma(1-e)cs=[204.5*0.2379/(165*(1-.454)*0.25)]$$

$$v=2.16 \text{ ft/hr}$$

Movement of 67.46(Tg@ 1hr.)=vt=2.16*(1hr)=2.16 ft

Location of point at start of heating=3.00-2.16=0.84 ft

$$h'=204.5*0.2378/[0.84*(1-.454)]Ln[(67.46-66)/(200-66)]$$

$$h'=479.4 \text{ btu/hr-ft3-'F}$$

Precision of this result is not as high as this procedure has capability to obtain. It is obvious that selecting an initial temperature of 67.46 leaves room for error when subtracted by

66'F. Selecting a higher temperature would have introduced increasing heat losses as shown by the 4 hrs required to reach 196'F. It can be shown that the hot gas had sufficient energy to heat the bed in 1.31 hrs. A better way would be to use a shorter bed length so a portion of the hot gas profile would have emerged in a time measured in seconds. Heat loss would not have had time to occur and the emerging temperature would have changed so slowly that a quality measurement of temperature could be taken. It is always true that bed housing will have heat capacity and will withdraw heat from the stream. Such losses should best be accounted separately and not lumped with a natural parameter. Equation 5 calculated h' as 419 for Run 10 and Lof & Hawley used Schumann's Curves to estimate a value of 437. The above value was 479. All would give reasonable bed lengths. Lower values of h' give longer S-Curves and longer increments to add to bed length.

Think of Equation 1 as the plug flow part of the solution. The term "plug flow part" is the recognition that Equation 1 represents that if the bed were to heat by plug flow then the extent of heating in the bed would be exactly as predicted by the product vθ. Because the hot end of the bed is represented by a symmetrical S-Curve, Equation 1 will represent the distance where 50%, or midpoint, of the temperature range is located.

Comparison to Schumann Solution

It came as a surprise to notice that Equation 4 without the x increment could be rearranged to give the parameters Y and Z. It was later in this study that the relationship of the Equations to Schumann's curves was first noticed. Neglect the insignificant second part of the denominator in Equation 1.

$$L= v\theta =\{Gcg/[\sigma cs(1-e)]\}\theta$$

If both sides are multiplied through by h' and divided through by Gcg a new equality is obtained.

$$Lh'/Gcg=(h'\theta/(\sigma cs(1-e)$$

These terms are exact duplicates of the terms Y and Z from Schumann's Curves. The relationship becomes clearer when the point where Y=Z is examined on the Curves. In each case the parameter

$$(Tg-Tso)/(Tgo-Tso)\cong 0.55\text{-}0.58$$

Schumann adds more length to beds with exit temperature lower than that ratio and subtracts length from beds with larger ratios.

It is obvious that Schumann's solution to the basic bed length prediction was quite accurate. The method selected for presenting the information to the Scientific Community was confusing. It has remained quite mysterious to some of us. His choice for placement of the gas S-Curve might be a useful study. Did he predict that the shape and length of the S-Curve would change as it traveled? Why did the slopes change as Z increased? Perhaps the variance was due to incomplete S-Curves at the start. The conclusion from this analysis is that Schumann discovered the principles discussed in this Paper but the simplicity of the solution was obscured by his Curves.

Calculating the Bed S-Curve

Some designers might prefer rigor to simplicity in bed design. For those, an equation has been developed for calculating the actual temperature profile for the bed. While analyzing the results from numerical calculations a technique was discovered that allowed the bed profile to be constructed. It required the addition of two exponential curves. One curve was the bed profile that resulted from allowing the temperatures on the bed face to move into the bed at the TR velocity. When area under that profile failed to contain all the heat supplied by the gas a second equation was discovered. It represented heat carried into the bed at temperatures higher than the momentary face temperature. After numerical construction of the beds temperature profile it became apparent that the two equations could be combined into one equation. Equation 9 shows the bed temperature verses distance along the bed. The bed profile is shown in Figure 2.

Equation 9 Cold Bed $x=v\{\theta-(\sigma cs/h')Ln[(Ts-Tso)/(Tgo-Ts)]\}$

Hot Bed $x=v\{\theta-(\sigma cs/h')Ln[(Tso-Ts)/(Ts-Tgo)]\}$

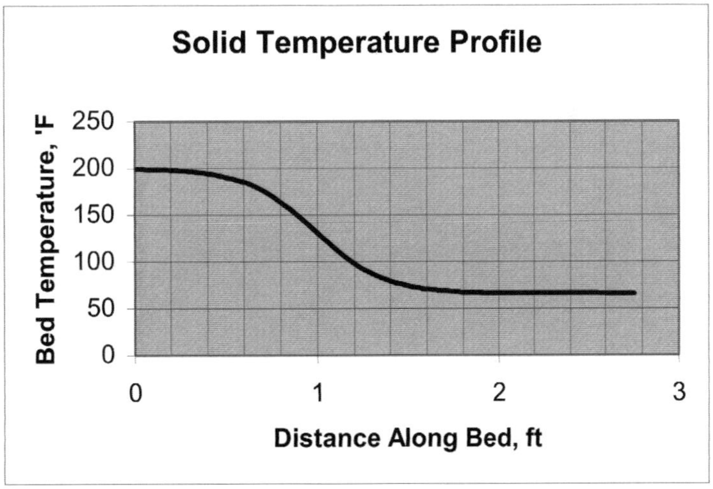

Figure 2. Developed from Lof &Hawley Data

Note that the initial slope of the S-Curve part is regulated by bed density, heat capacity, the heat transfer coefficient and initial bed temperature profile. In verbal communications with Loren Hov, P.E., Reference 11, former Director of Energy Management of the former Stauffer Chemical Co., now part of Syngenta, who proof read the paper offered indisputable evidence that S-Curves flatten in slope each time reverse flow enters the bed until an equilibrium is established. There is little doubt that the initial gas temperature profile dictates the initial S-Curve slope. An existing S-Curve will increase the length of the cooling profile and add length to the S-Curve. Design and operation of Regenerative beds must overcome that difficulty if the initial steep S-Curve is to remain.

Air and Bed Profile

The air temperature profile can be described in a logical way by moving the bed profile along an additional distance. Figure 3 shows the double profile. Correctly locating the air profile is done by trial. The bed profile is moved forward a selected distance by making θ equal to $\theta+\Delta\theta$. A numerical integral is computed using Equation 10 along the Double S-Curves where values of Tg-Ts are at equal x values.

Equation 10 $$Q=\text{Integral}\,[h'(1-e)(Tg-Ts)dx]=Gcg(Tgo-Tso)$$

Based on the outcome of trials the air profile is moved until Equation 10 is correct. Values of Tg-Ts can be scaled from the graph. If it is recognized that the points of interest on the two S-Curves fall where values of x are equal then a helpful expression can be derived. Equation 11 resulted from setting common values of x equal and solving for the difference in temperature between bed and fluid. Values for the temperature difference are then placed in the numerical integral table. Next the numerical integral is developed for (Tg-Ts)dx. Once the integral has been established and converted to total heat transferred then simply adjust $\Delta\theta$ until the total heat is correct. Equation 11 calculates values of (Tg-Ts) for selected values of Ts.

Equation 11 $$Tg-Ts=(Ts-Tso)[(e^\wedge\alpha)-1]/(1+[(Ts-Tso)/(Tg-Ts)]e^\wedge\alpha)$$

$\alpha=h'\Delta\theta/(\sigma cs)$
e is base of Natural Logarithm only in Eq. 11.

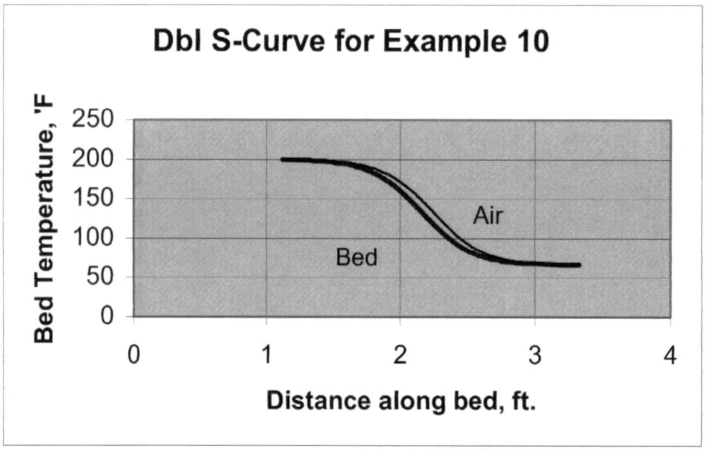

Figure 3. Air Bed Profile from Example 2

Conclusions

Equation 9 allows an accurate bed temperature profile to be drawn and Equation 10 allows the air profile to be placed correctly. However, there is little if any improvement over the method applied to Example 1 unless a fluid exit temperature higher than 50% of range is selected. Equations 1 to 5 will provide all the accuracy for bed design that will normally be needed. The simplicity associated with the latter equations will allow quick designs to be developed for streams that were of no commercial value prior to the rise in energy cost. Selection of equations will be the choice of the reader. Equations in this Paper will also find useful application with Rotary Kilns. The equations will not perform under a reverse flow arrangement where the new flow makes contact with and existing S-Curve. Present designers who apply reverse flow in bed designs must operate with long S-Curves and short cycles. Reverse flow greatly simplifies the materials of construction but sacrifices bed capacity. That type Regenerator is described in the text “Heat Transfer” by Max Jacobs (Reference 12).

References

1. Dabora, E. K., Moyle, M. P. ET AL; Description and Experimental Results of Two Regenerative Heat Exchangers; AIChE Symposium Series, No. 29, Vol 55.
2. Schumann, T.E.W., J. Franklin Inst., 208, 405 (1929).
3. Kern, D.O. Process Heat Transfer, McGraw-Hill, pages 668-671 (1950)
4. Thompson, R. J., Jacobs, H. R. AIChE, Symposium Series, No 236, Vol 80.
5. Balakrishnan, A. R. & Pei, D. C. T., Ind. Eng. Chem., Process Des. Del., 18(1), 30 (1979)
6. Lof, G. O. G. & Hawley, R. W., Ind. & Eng. Chem., Vol. 40, No. 6 (1948)
7. Furnas, C. C., Trans. Am. Inst. Chem. Engrs., 24, 142 (1930)
8. Furnas, C. C., U. S. Bur. Mines Bull. 361, (1932)
9. Saunders, O. A., & Ford, H., J., Iron Steel Inst., (London) 1, 291 (1940).
10. Foust, A. S., Wenzel, L. A., ET AL, Principles of Unit Operations, John Wiley & Sons, Inc. (1960)
11. Hov, Loren, P.E., Consultant, 1212 58th Ave., Sacramento, CA, 95831, Phone 916 391 5855, Mr. Hov in the 1950's designed, built, and ran duplex large pebble bed heat exchangers to not only exchange heat but to freeze reactant gas compositions where the temperature differential in and out was over 1600'F. Owner Stauffer-Aerojet Chemical Company.
12. Jacobs, M., Heat Transfer, Vol 2, John Wiley and Sons, pages 269-311. (1964)

Application of Waste Heat Recovery in Glass Industry

Robert Eberts, C.E.M.
Shell Services USA
Pennzoil North Tower
700 Milam Suite 15044
Houston, TX 77002

Introduction

To date the majority of innovation around energy efficiency in the fibrous glass industry has focused on furnace and melting operations. However, recent environmental regulations have increased the energy intensity of post forming operations through specific emission controls. This paper covers the application of waste heat recovery from incinerators operated in the fibrous glass industry.

The drying and curing of thermosetting binder solution used in production of wool insulation produces phenolic gases. Environmental regulations dictate that these harmful gases must be incinerated before effluent from the curing ovens may be exhausted into the atmosphere. This proposal is for the installation of multiple heat exchangers in the exhaust gas streams of incinerator units at a fiberglass production facility. The waste heat recovered would then be used to preheat both the inlet stream to the incinerator and the combustion air to the associated curing oven.

Process Description

In the final step of the forming process, glass wool fibers are sprayed with a binder solution and collected on a conveyor belt. The mat is carried through a large oven where the thermosetting binder is cured (see Figure 1). Depending on the thickness of the mat, curing temperatures can range from 350°F to 900°F. Gaseous volatile organic chemicals (VOC) are produced during curing. The 1990 Clean Air Act regulates the release of these gases. The preferred method of control is thermal incineration.

Table 1: Emission Factors for Rotary Spin Fibrous Glass Post Forming [1]

Filterable PM	NO_x	CO_2	VOC	Formaldehyde
6	2	3.5	7	0.994

Notes: Units are lbs per ton of finished product

Exhaust hoods are located on both charge and discharge ends of the oven to collect the oven effluent for thermal incineration. Often, gas-fired burners are located in these exhaust ducts to prevent cooling of the gas which leads to condensing of the vaporized binder. Emissions control is achieved through maintaining a specified outlet temperature from the thermal incinerator.

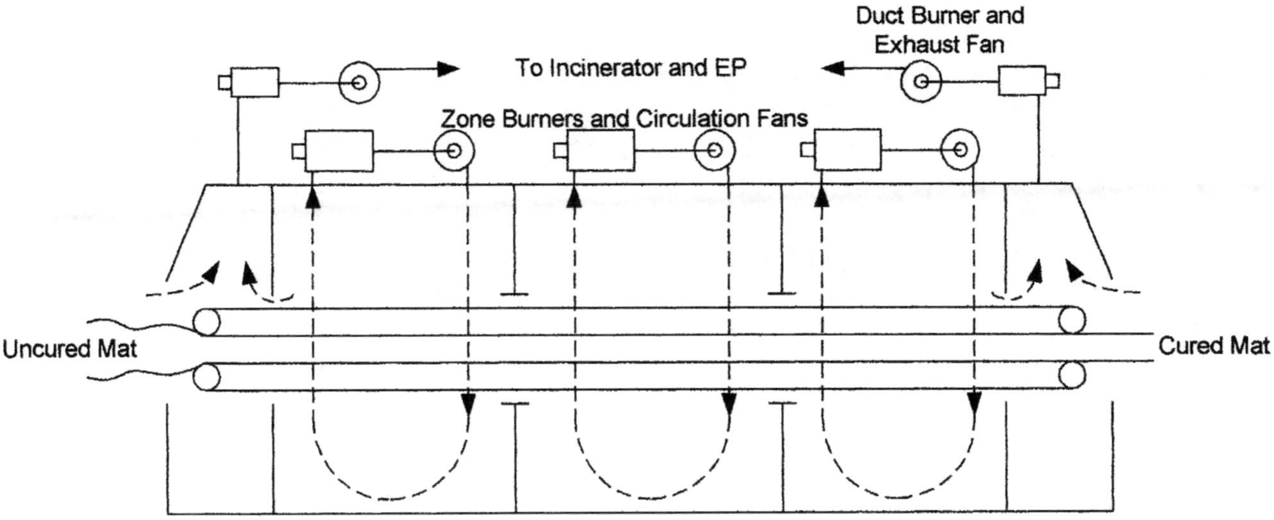

Figure 1: Wool Insulation Curing Oven Schematic

After the curing oven, the mat passes through a cooling section where ambient air flows through the mat. If hot spots are left in the mat these can ignite and spread. Uneven edges are trimmed from the mat. Facing may be applied to the mat using an adhesive. Once the mat is trimmed and finished, it is cut to specification and packaged.

Considerations

According to a 1988 study conducted by Energetics for the US Department of Energy, "Fiber glass wool curing consumes about 470 to 588 Btu/lb of glass; incineration of phenolic fumes consumes about 588 Btu/lb." [1] However, depending on size, age, and condition of the equipment these numbers can be more than double these theoretical values.

Analysis of two separate curing ovens was conducted and found to have considerably higher thermal requirements as outlined in Table 2.

Table 2: Observed Post Forming Energy Intensity

Thermal Load	Oven A	Oven B
Curing of thermosetting binder	2,168	1,376
Incineration of oven effluent	1,639	1,203
Total Thermal Load	3,807	2,399

Notes: Values given as Btu per lb of finished product.

Opportunity for waste heat recovery was very apparent. Several other energy optimization opportunities were identified and implemented reducing fuels consumption. However, the thermal efficiency of the curing ovens remained quite low compared to theoretical values. Environmental permit regulations stipulated that the outlet temperature of the thermal incinerators could not drop below 1,200°F. Standard operation maintained a safety margin above this minimum of 10-15°F. This high temperature gas stream could be recovered to pre-heat several gas streams in the vicinity. Minimizing equipment and installation costs.

Design

The proposed waste heat recovery system focused on two potential gas streams. First, the oven exhaust would be pre-heated in two stages before entering the thermal incinerator lowering fuel consumption at the incinerator. Second, the incinerator exhaust stream would pass through a third gas-gas heat exchanger pre-heating combustion air to the curing oven effectively lowering the fuel consumption of those units as well. The two systems, A and B, are very similar in concept and application. In addition to the cross flow heat exchangers; each system would require the purchase of additional fans to handle the pressure drop created by the heat exchangers and instrumentation for monitoring and control. Figure 1 shows a simplified schematic of the oven exhaust system with waste heat recovery.

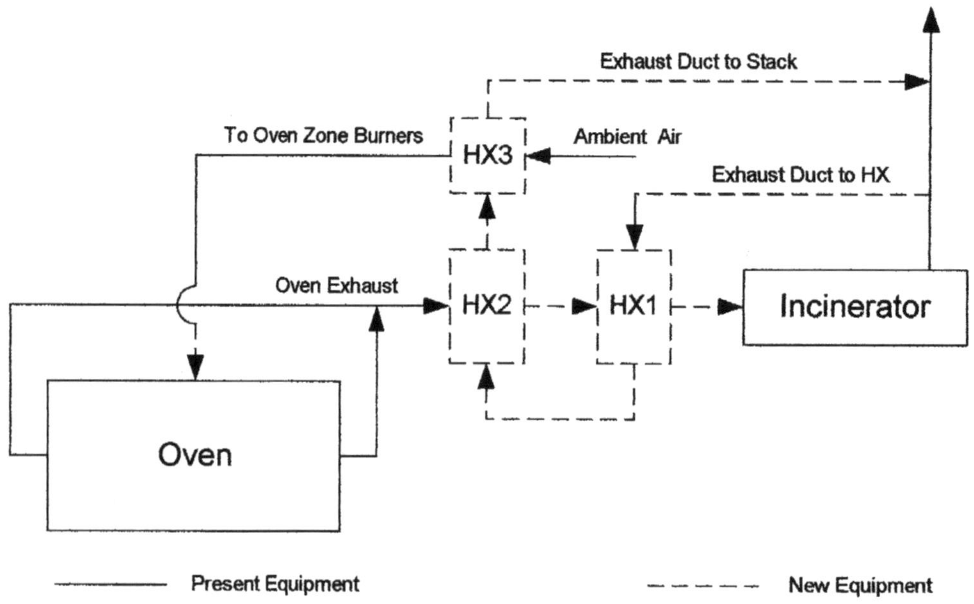

Figure 2: Heat Recovery Schematic

Key design parameters for the two systems are outlined in the Table 3.

Table 3: Waste Heat Recovery Design Parameters

Design Parameter	System A	System B
Oven Combustion Air Flow [lbs/hr]	11,420	11,612
HX3 Cold Gas Outlet Temp. [^oF]	400	400
Oven Exhaust flow [lb/hr]	78,750	96,750
Oven Exhaust temp [^oF]	500	500
HX2 Cold Gas Outlet Temp. [^oF]	635	635
HX1 Cold Gas Outlet Temp. [^oF]	1,000	1,000
Incinerator Outlet Temp. [^oF]	1,200	1,200
HX1 Hot Gas Outlet Temp. [^oF]	850	850
HX2 Hot Gas Outlet Temp. [^oF]	625	625

HX3 Hot Gas Outlet Temp. [°F]	311	311

In addition to the purchase of six (6) heat exchangers, five (5) fans, the associated ductwork, and electrical drives much is required for the operation of the waste heat recovery system. Instrumentation for monitoring and control includes automated dampers, pressure indicators/transmitters, temperature indicators/transmitters, associated wiring, and computer hardware. Structural and civil estimates include allowances for platforms, supports, and site preparation.

Financials

Estimates on the completed proposal revealed a potential savings of 28MMBTU per hour of operation or approximately $1.3MM annually. The majority of this savings, 23MMBTU per hour, is directly related to waste heat recovered by the heat exchangers. An additional 5MMBTU per hour is saved indirectly through a reduction in combustion air required at the oven zone burners and incinerators. An economic feasibility study was conducted and resulted in an estimated installed cost of $1.6MM. The simple pay back period of less than 15 months make this a viable project for further investigation.

In summary, this case for waste heat recovery is very strong. The heat content and quality of the recovered air stream make it inherently valuable. Close proximity of heat sinks, which directly impact the thermal load of the facility, minimizes installation costs while maximizing savings. Proven methods and technologies assure stable operation for many years. Even in this time of reduced capital investment this project should be very competitive for funding.

References

1. *Energy and Environmental Profile of the U.S. Glass Industry*, U.S. Department of Energy, Office of Industrial Technologies, April 2002.

Paper 40c

PX PROCESSES THAT UTILIZE WASTE HEAT POWERED AMMONIA ABSORPTION REFRIGERATION

Jeffrey A. Amelse ** and Phillip L. Jackson

BP Amoco Chemical Company
150 West Warrenville Road
Naperville , IL 60566
AmelseJA@BP.com

** To whom correspondence should be sent

D. Erickson, C.B. Panchal, G. Anand, and E. Makar
Energy Concepts Company
627 Ridgely Ave.
Annapolis, MD 21401

K.D. Ganesan and A. Cato
Fluor Corporation
100 Fluor Daniel Drive
Greenville, SC 29607

Keywords: ammonia, absorption, refrigeration, paraxylene, waste, heat

Prepared for Presentation at the 2004 Spring National Meeting, New Orleans, LA, April 25 – 29
UNPUBLISHED

This project is supported by the U.S. Department of Energy (DOE) Award Number DE-FC36-03ID14442. However, any opinions, findings, conclusions, or recommendations expressed herein are those of the author(s) and do not necessarily reflect the views of the DOE.

Abstract

Amoco Chemical Company and now bp Amoco Chemical Company (bp) have continued to improve their xylene isomerization catalysts and multi-stage pX crystallization processes since commissioning their first commercial pX unit in 1967. pX separation from isomerized reformate mixed xylenes feedstocks requires first stage crystallization section temperatures as low as –68 °C (-90 °F). The current bp processes use cascaded ethylene/propane or ethylene/propylene (C2/C3) vapor compression refrigeration systems. The C3 compressor is an expensive equipment item, and a large user of electricity. This paper describes new energy efficient pX processes that use waste heat powered ammonia absorption refrigeration (AAR) instead of conventional C3 refrigeration. A 400 KMTA pX unit based on reformate xylenes that uses AAR instead of C3 refrigeration will save on the order of 4.9 MW of electricity. This corresponds to an indirect reduction in CO_2 green house gas production of about 40 KMTA.

Background

Para-xylene (pX) is a bulk industrial chemical intermediate. It is oxidized and purified to Purified Terephthalic Acid (PTA). PTA is reacted with ethylene glycol to produce polyester, which has a number of uses, including fiber for clothing, film, tire cord, and plastic beverage bottles. World capacity for pX was about 24 million metric tons per annum in 2003. bp Amoco Chemical Company (bp) is one of the largest producers of pX and PTA.

The primary feedstock for pX is reformate mixed xylenes. Reformate mixed xylenes account for about 80% of mixed xylenes production. Reformate mixed xylenes are separated from refinery reformate by distillation. This fraction is comprised primarily of the C8 aromatics (ethylbenzene (EB), and the three xylene isomers (ortho-, meta-, and para-)). The equilibrium distribution of the xylenes is approximately 25% pX, 50% mX, and 25% oX. The pX concentration in the feed is further diluted by the presence of about 12-20% EB, and by up to several weight percent non-aromatics. The C8 aromatics have very close boiling points, and cannot be separated by distillation. Thus, pX is separated from mixed xylenes by either low temperature crystallization or by selective adsorption on a molecular sieve. bp uses its own pX crystallization process.

A generic pX process based on reformate xylenes is comprised of three sections: isomerization, fractionation, and pX separation, as shown in Figure 1.

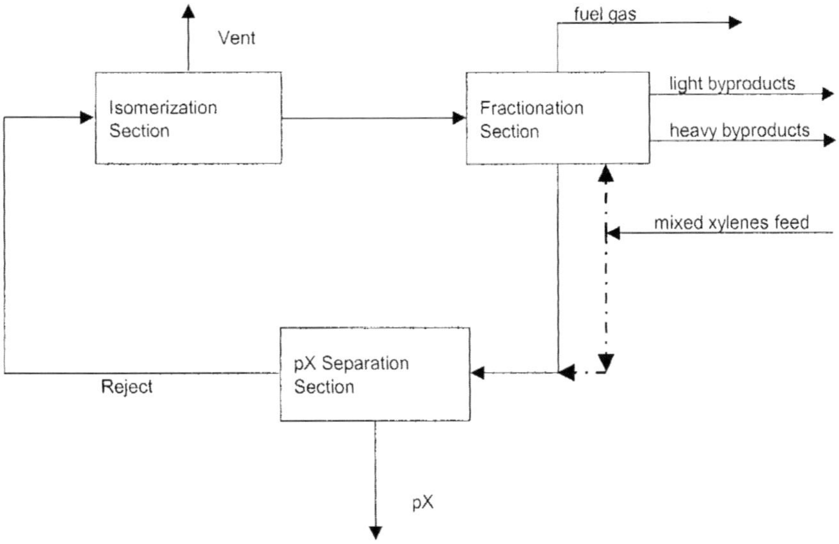

Figure 1. Generic pX Process

The pX lean reject from the separation section is isomerized to re-establish the equilibrium distribution of the xylenes. The isomerization section catalyst must also convert EB and non-aromatic byproducts to lighter or heavier byproducts that can be readily removed in the fractionation section to prevent their build-up in the recycle loop.

pX crystallization sections are generally comprised of several stages. Low pX concentration (due to dilution by EB, non-aromatics, and the other xylene isomers) and high pX solubility in the other isomers requires first stage temperatures as low as –68 °C (-90 °F). To achieve these low temperatures, current bp pX crystallization processes use cascaded ethylene/propane or ethylene/propylene (C2/C3) vapor compression refrigeration systems, as shown in Figure 2.

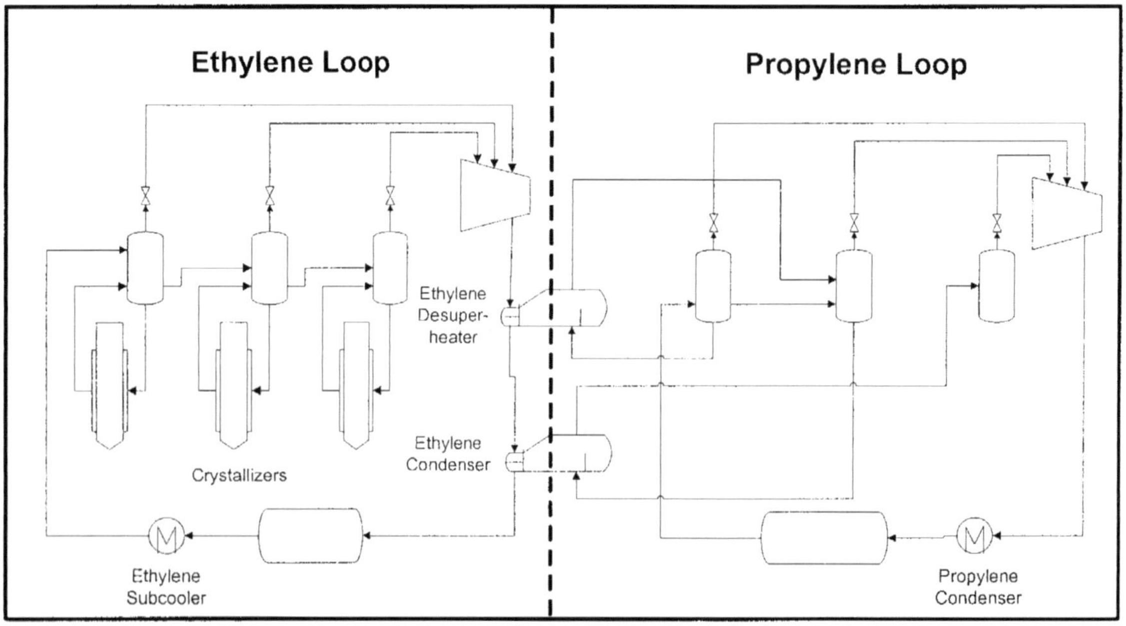

Figure 2. Cascaded Ethylene/Propylene Refrigeration System

The ethylene refrigerant is used to achieve the very low first stage temperatures. In vapor compression refrigeration cycles, heat is removed from the process by boiling a liquid refrigerant at low pressure, and thus, low temperature. The vapor is compressed to higher pressure at which the heat of vaporization can be removed at a convenient higher temperature. The critical temperature of ethylene (about 9.5 °C) is below ambient. Thus, ethylene cannot be condensed in an air or water cooler. Rather, ethylene is condensed as it boils a liquid C3 refrigerant. The C3 refrigerant vapor is then compressed and condensed in an air or water cooler. Liquid C3 refrigerant can also be used to remove heat from warmer second or subsequent crystallization steps.

The C3 compressor is an expensive equipment item, and a large user of electricity. This paper describes pX processes that use AAR to replace the C3 refrigeration loop. Instead of electricity, AAR is powered by waste heat available on the pX unit or an associated feedstock preparation unit that isolates mixed xylenes feed from refinery reformate.

The waste heat that powers AAR is low temperature heat that is otherwise lost to the environment via air or water coolers. Thus, it is essentially free energy. A simple single-effect AAR cycle is shown in Figure 3.

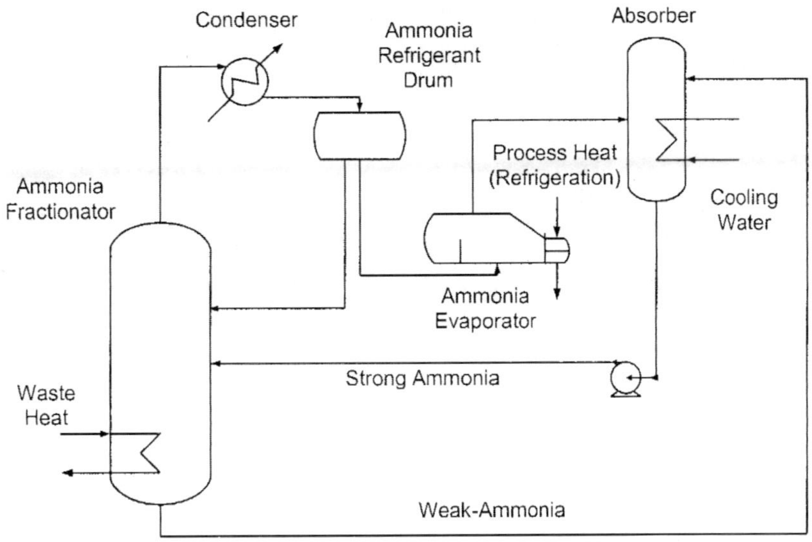

Figure 3. Single-Effect AAR Cycle

As in conventional C3 refrigeration, the AAR process removes heat from the process by boiling a liquid refrigerant (in this case liquid ammonia instead of C3). In conventional C3 refrigeration, the refrigerant vapor is compressed to a high enough pressure that it can be condensed against air or water. In the AAR process, the ammonia vapor is absorbed in an aqueous solution. Absorption is exothermic and heat is removed from the absorber by cooling water. The waste heat is used to distill the absorber solution into essentially pure liquid ammonia refrigerant and a very dilute ammonia solution, which is recirculated to the absorber. AAR can achieve temperatures in the range of those achieved by C3 refrigeration, and modern multi-stage advanced AAR cycles are capital cost competitive with conventional C3 refrigeration.

The waste heat that drives the AAR process replaces electricity used to run the compressors of conventional C3 refrigeration systems. Thus, the new pX/AAR processes are GREEN in that they will save a considerable amount of electricity. Since incremental electricity comes from coal-fired power plants, a reduction in electricity translates to an indirect reduction in CO_2 green house gas.

Mixed xylenes feed can also be produced by toluene disproportionation (TDP). In TDP, two moles of toluene react to form one mole of xylenes and one mole of benzene. Selective TDP (STDP) processes produce mixed xylenes with a pX content that can exceed 90%, well above equilibrium (Young, et al.; 1982). STDP does this via the use of shape selective molecular sieves. The reaction is limited by product diffusion rates. pX diffuses through the molecular sieve 10-1000 times faster than the other isomers.

pX can be recovered by crystallization from high pX concentration STDP xylenes at relatively warm temperatures that can be achieved by C3 refrigeration. AAR can be used instead of C3 refrigeration to supply the bulk or all of the refrigeration duty needed to recover pX from STDP xylenes. By avoiding the deep ethylene refrigeration, STDP pX units require much less capital and energy than pX units that use reformate xylenes feed. Thus, STDP processes have grown in popularity, and accounted for about 6% of pX produced in 2003.

Structure of the DOE Project

The U.S. Department of Energy (DOE) is supporting bp, Energy Concepts Company, and Fluor Daniel in the development of pX processes that use waste heat powered AAR via its cost shared Chemical Industry of the Future program.

This project is structured into three phases spread over three years:

Phase 1: Proof of Concept and Economic Incentive

Phase 2: Optimize pX/AAR Designs

Phase 3: Construct a Commercial Scale Demonstration Unit.

Phase 1

Phase 1, completed in October, 2003, involved demonstrating proof of concept and economic incentive with minimal effort. AAR units were designed as drop-in replacements for the C3 systems of recent bp crystallization units. Two cases were considered: reformate xylenes and high pX (STDP) xylenes. To minimize effort, the pX unit isomerization, fractionation, and crystallization sections were frozen. Only the C3 refrigeration section was modified (replaced by AAR).

Current bp processes for recovering pX from reformate xylenes employ high heat integration between the isomerization and fractionation sections to reduce fuel and capital. In fact, the current bp processes are so fuel efficient that without rearranging the heat integration, there is not enough heat available on the pX unit to power the AAR. Thus, the Phase 1 reformate xylenes case AAR drew about one third of its heat from the pX unit, and about two thirds of its heat from a mixed xylenes feedstock preparation unit (two distillation towers used to isolate mixed xylenes from refinery reformate), and the pX unit became dependent on a co-located feedstock preparation unit. Nevertheless, this case demonstrated proof of concept for the reformate xylenes case.

To prove the concept, an advanced AAR cycle was developed that was able to make use of the available waste heat sources and deliver the required refrigeration duties at the required ammonia evaporation temperatures. The lowest ammonia evaporation temperature encountered is that required to condense ethylene (about –37 °C or –35 °F). The pX unit heat source was a sensible heat stream that glides from about 139 °C to about 92 °C (282 °F to 197 °F). The other heat source was condensing overhead vapor of the detoluenizer in the feedstock preparation unit that glides from about 135 °C to about 125 °C (274 °F to 257 °F). The feedstock unit is comprised of two distillation towers that isolate mixed xylenes feedstock from reformate. To make use of these two heat sources and to deliver two temperature levels of liquid ammonia refrigerant, an advanced multi-effect cycle was chosen, which is a slight variant of the cycle shown in Figure 4.

This advanced multi-effect AAR cycle is analogous to the multi-stage vapor compression cycles shown in Figure 2. It is comprised of a single-effect section which supplies a –7.8 °C (18 °F) chilling (IP evaporator) plus a double lift section, which supplies a –37.2 °C (-35 °F) chilling (LP evaporator). Each section has an associated refrigerant heat exchanger (RHX) and absorber, and they both share the high pressure rectifier and reboiler. The double lift section additionally incorporates a vapor exchange (VX) rectifier, reboiler, and absorber. In some instances the IP absorber and VX absorbers can be combined, but in this flow sheet the available heat source temperatures did not permit this. This cycle employs several Energy Concepts Company patented proprietary technologies, including the "vapor exchange" cycle, a patent pending technique for using waste heat which is at a lower temperature than the column reboiler, and compact, high performance heat and mass exchangers (U.S. Patent 5,713,216; US 5,798,086; US 6,203,718; US 6,584,801B1; US 6,585,238B1; and US 6,679,083B1).

The Phase 1 reformate case AAR design achieves a Coefficient of Performance (COP) of 0.349, where COP is defined as:

COP = (refrigeration duty delivered)/(waste heat input)

The Phase 1 reformate xylenes case design would save about 4.9 MW per 400 KMTA (thousand metric tons per annum) pX unit. This translates into an indirect reduction in CO_2 from coal-fired power plants of about 40 KMTA. The electricity cost savings and environmental benefits alone provided strong incentive to proceed with AAR. However, The AAR will also have a lower installed capital cost.

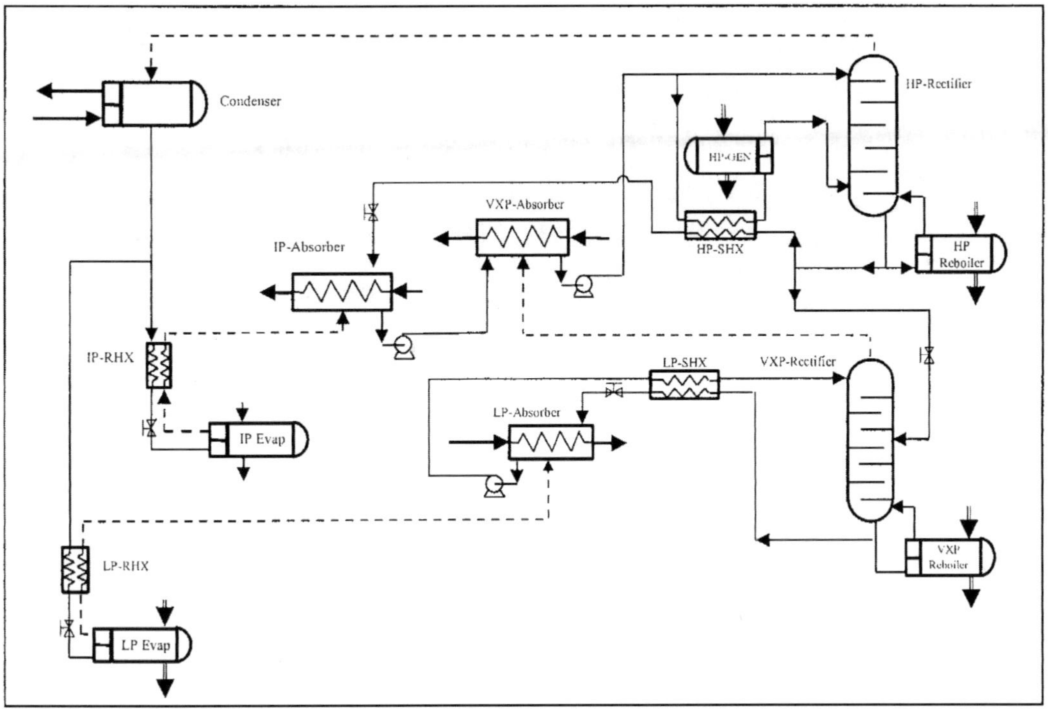

Figure 4. Advanced Multi-Effect AAR Cycle

The Phase 1 STDP case indicated that there is more than enough heat available from the condensing overhead vapors of the STDP unit hot end fractionation towers. STDP hot ends typically operate below equilibrium toluene conversion (which itself is only in the range of 35% TOL conversion). Thus, the towers required to isolate the STDP mixed xylenes product from recycle toluene, benzene, and heavy byproducts are large sources of waste heat. The heat is available at higher temperature, and with less glide. This leads to a slightly less complex AAR cycle than for the reformate xylenes case, smaller AAR equipment, and even higher capital savings. The Phase 1 STDP AAR design achieves has a COP of 0.396.

The economic incentives demonstrated in Phase 1 justified continuation to Phase 2.

Phase 2

Phase 2 also considers two cases: reformate mixed xylenes and STDP xylenes feedstocks.

The goal for the Phase 2 reformate xylenes case is to shift all of the waste heat source onto the pX unit, to make the pX unit independent of the feedstock unit. By rearranging the pX unit heat integration, a preliminary Phase 2 reformate xylenes case has been developed that achieves this goal without increasing pX unit fuel consumption! About 85% of the waste heat comes from a pX unit hot stream, and the balance comes from steam raised in the convection sections of the isomerization and fractionation section furnaces. The current bp pX process does not use much steam, and thus, steam is not generated in the furnaces. By adding steam generation to the furnaces for the pX/AAR design, the overall furnace efficiency is increased from about 80% to about 90%, an environmental bonus!

bp owns and operates an STDP unit at its Texas City site. The pX crystallization recovery section is of Amoco (now bp) design. The original design incorporated a small ethylene refrigerated scavenger stage to minimize reject xylenes. The goal for the Phase 2 STDP xylenes case is to eliminate this ethylene-refrigerated stage, and to use only AAR refrigeration in place of C3 refrigeration. Eliminating the ethylene stage will eliminate a crystallizer, centrifuges, the ethylene compressor, and associated equipment, which will lead to large capital and energy savings!

Conclusions

Waste heat powered ammonia absorption refrigeration is an attractive replacement for the C3 refrigeration section of the cascaded C2/C3 refrigeration systems used in current bp pX crystallization units. After rearranging heat integration, ample waste heat is available on the pX units or associated feedstock units to power the AAR. Thus, high electricity consumption by the C3 refrigeration compressors can be replaced by "free" waste heat. Electricity savings will be on the order of 4.9 MW per 400 KMTA pX unit based on reformate xylenes. This corresponds to an indirect reduction in CO_2 from coal-fired power plants of about 40 KMTA. As an added bonus, advanced cycle multi-effect AAR units have a lower capital cost than comparable C3 vapor compression refrigeration systems.

Acknowledgement

This project is supported by the U.S. Department of Energy (DOE) Award Number DE-FC36-03ID14442. However, any opinions, findings, conclusions, or recommendations expressed herein are those of the author(s) and do not necessarily reflect the views of the DOE.

REFERENCES

Young, L. B., Butter, S. A., and Kaeding, W. W., J. Catal., 76, 418 (1982).

Paper 41b

STEAM COST INFLUENCE ON ENERGY EFFICIENCY

Jerry L Robertson, Fellow AIChE
Retired from Exxon Research and Engineering Co.
Jlrobert20@charter.net

Key Words: energy efficiency, steam, refinery, petrochemical

Prepared for Presentation at the 2004 National Meeting, New Orleans, LA, April 25-29, 2004

Unpublished

Steam Cost Influence on Energy Efficiency

Jerry L Robertson, Fellow AIChE
Retired from Exxon Research and Engineering Co.
Jlrobert20@charter.net

Introduction

Steam is an often unappreciated but essential ingredient for operations in refineries and chemical plants. Because it is centrally generated or purchased, it often managed by a part of the organization other than the part or parts that manage each process sector. Steam costs and credits can have a major impact on project economics as well as day to day operating optimization. There have been many efforts to calculate the **cost** of steam some of which involve the integration of mechanical work demands with the heat requirements of the plant. The focus of this paper is a method that calculates the **value** of steam to the process areas as it depends on process conditions and ratio of the cost of fuel to the cost of heat exchangers. This focus automatically leads to an emphasis on temperature (rather than pressure) and heat exchangers (rather than turbines).

Uses and value of steam

Because of the many and sometimes unrelated uses of steam at more than one pressure (temperature) steam systems are both dynamic and complex. Some steam loads such as stripping or reboiler applications are nominally consistent with time. Others such as flare mixing steam or furnace decoking steam are off/on. The ability to deal with the dynamics in demand while maintaining a consistency in quality and reliability affects the characteristics and thus the cost of the steam system.

Normal Steam User	Alternative to Steam	Advantages (Disadvantages)
Light ends stripping	Inert gas Reboiler vapor	Water--ease of recovery. Steam reduces partial pressure of hydrocarbons
Steam Turbines	Electric Motors	Reliability, Variable speed (Steam piping maintenance) (Condensate recovery)
Heat Tracing	Electrical Tracing	Newer Technology (Steam trap maintenance)
Fluidization Steam	None proven	
Reboiling Light Ends	Direct heat integration	Higher heat transfer coefficient Better control
Reboiling Acid Gas Regeneration Towers	Direct heat integration	Increased direct heat integration can free up steam for this service
Vacuum ejectors	Vacuum Pumps	Reliability
Hot belt to transport heat --possibly several steam levels--	Direct heat recovery Hot oil belts	High heat transfer coefficient Convenient piping. Fungible (Reduce □T), (Must be saturated)

Table 1 Steady State Steam Users and Alternatives

Steam demands are also intermittent. It is used as a mixing force to induce additional air into smoke-free flares when upsets require flaring; and as a decoking driver for some furnace tubes. "Fluidization" steam in case of feed or air blower outage can result in large instantaneous demands. It is used as an emergency cooler when the internal insulation fails on a very hot reactor.

Because of these intermittent requirements, steam systems have a capacity significantly higher than average operation which often causes non optimum boiler operation. A cold rainstorm can rapidly add to the steam system demand and cause significant upsets. The larger the system with greater quantities of small piping the greater this rain load can be.

This scope and complexity makes analysis of the marginal and all in economics of the steam system a challenge. With the exception to fluidization steam, each of these uses has alternatives. Each of the alternatives to steam usage can be evaluated based on the **value** to the system of utilizing that alternative.

The value is the reduction in total cost for the entire system that occurs as a result of the implementation of the alternative. For example the value of an air mixer for a flare might be the cost savings that result from the shut down of a boiler that is "simmering" in stand by waiting for the next event that requires flaring minus the installed cost of the air mixing system. A simple method for approximate evaluation of the value of steam heating as an alternative for process heat integration is valuable as a tool for steam system analysis.

Cost of Steam

Knowing the true cost of steam is vitally important to determining the true value to the system of its use. Whether it is the "all in" or average (including capita)l cost of all steam produced or just the marginal cost of only the steam "saved" as the result of the substitution of one of the alternatives mentioned in Table 1, will depend on the situation. For existing steam systems, the cost impact of a project is usually marginal since it is usually impossible to recover any capital costs from an existing steam system. The marginal cost usually covers the cost of fuel, the operating cost for water pumping, and the cost of chemicals and includes factors for boiler efficiency. For waste heat recovery, the cost of fuel is eliminated.

Steam costs as a function of pressure have been based on enthalpy or First Law (for systems that involve only waste heat boilers and fired boilers) or availability or Second Law for systems that involve both mechanical work from turbines and heat input into the system[1]. Ultimately, a model of the entire steam system that includes many of its cost and process characteristics will provide the most reliable results[2]. While the tools to support building such a model have greatly improved, it remains a major effort. Cogeneration systems also add to the complexity to steam system analysis with the major question being whether full benefits of cogeneration go to heat or power[3]. Including the cogeneration system within the envelope of steam system analysis will provide the greatest amount of information on which to base decisions on expanding, modifying and operating a steam system.

Marginal Cost and Value of Steam for Heat Integration

Building a relatively simple heat exchange model allows calculation of value and cost of steam to a process system. It is based on the Incremental Value of Energy Saved (IVES) ($/BTU)which includes savings for fuel fired, cooling water saved and costs for energy consumed for power in heat transfer. It involves the Incremental Total Installed Heat Exchanger Cost (ITHEC) ($/ft^2) for the last square foot of heat exchanger surface added to the system. This includes the incremental cost of piping, foundations, etc. and labor to install the exchanger.

1. Define the Incremental Economic Heat Flux (IHF)[4] $= \dfrac{\text{ITHEC x Capital Recovery Factor}}{\text{IVES x Operating Hours per year}}$
2. Model the system to calculate the actual incremental heat flux based on the process conditions that establish the composite T-Q curves and the actual (or best calculated) heat transfer coefficients for the best arrangement of heat exchanger services. Call this the Heat Exchanger Pseudo model (HEP)
3. Calculate the furnace duty for the HEP for Incremental Heat Flux for the HEP that is equal to the Incremental Economic Heat Flux.
4. Model the system to make or use increments of steam and determine the furnace duty at the IHF set by economic conditions. The change in furnace duty will be the cost or value of the steam to the process system and represents the net system cost or savings per unit of energy input compared to a base case of no steam generation or use.

If the value is above the cost, it is a good project.

A Simple Heat Recovery Model

The following example is based on a simple heat exchanger model of the entire system using the now familiar "pinch" methodology with a few additional assumptions. For an entire process or refinery system, the Cold streams and Hot streams composite curves are approximated by three straight lines each. For cold stream, the lowest temperature zone represents heating liquids; the middle zone represents vaporizing (and heating) liquids; and the hottest Zone represents superheating vapors. For hot stream, the hottest zone represents cooling reactor product vapors; the middle zone represents condensing (and cooling) liquids; and the lowest zone represents cooling liquids and separating light products. Slopes and transition temperatures for both curves are set by process conditions—pressures, reflux ratios, reaction requirements, recycle rates, etc. The relative position of two curves is set by fired heater duty with increased overlap reducing both the fired heater duty and the cooling duty. The overlap of curves represents heat recovery and is modeled as several heat exchangers with heat assumed to be transferred vertically from the hot stream to the cold stream. Fixing the furnace duty, temperature and quantity of steam made and used fixes all the temperatures.

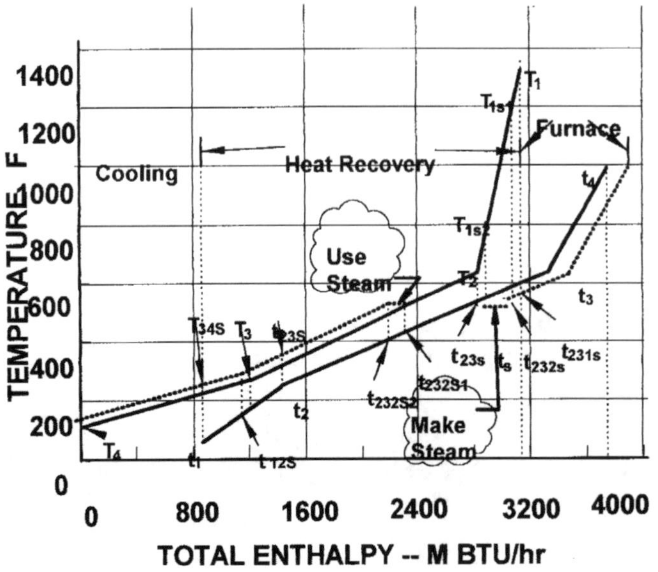

Figure 1 System Heat Exchanger Model

This model is based roughly on the Composite Curves for a low conversion 200,000 B/day refinery. Cold streams are represented by lower case and hot streams upper case with the numbering starting at the inlet of the system. Steam is made only with the hottest stream and used only in the middle zone in each case shifting the composite curves. If no steam is made or used, there are four zones for heat recovery with the hottest zone being zone 1.

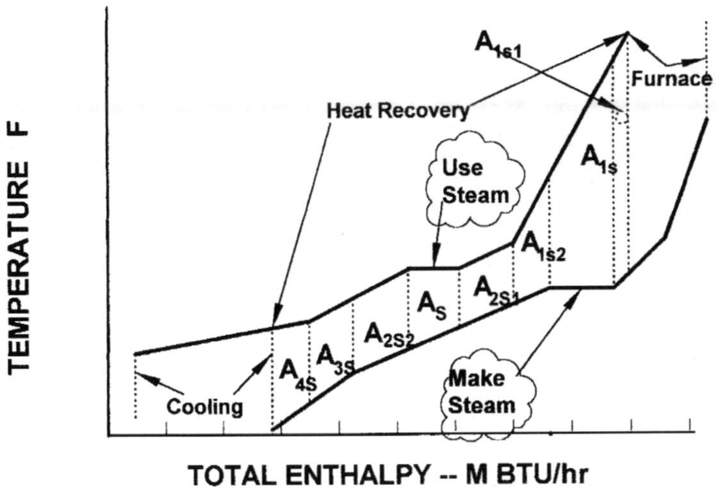

Figure 2 Area Model for Heat Recovery System
Heat is assumed to be transferred, from the hot stream at a temperature directly above the cold stream temperature in each area shown. This is the most efficient heat transfer and is a much more stringent assumption than "no heat transfer across the pinch". The area for each zone is calculated with a conventional countercurrent heat exchanger model. The furnace duty is perturbed and the area recalculated. The Incremental heat flux (□Q/□A) is calculated. The furnace duty is adjusted until the economic Incremental Heat Flux is reached.

Cold Streams		Hot Streams	
Temperature Fahrenheit	Enthalpy M BTU/hr	Temperature Fahrenheit	Enthalpy M BTU/hr
60	870	1350	3045
250	1410	630	2790
650	3340	270	1200
1000	3715	110	0
These were starting points before adjusting Furnace Duty			

Zone	Heat Transfer Coeff. BTU/hr-ft^2 F
A_1s_1	35
A_1s	60
A_1s_2	35
A_2S_1	30
A_S	100
A_2S_2	30
A_3S	35
A_4S	25

Table 2 Temperature-Enthalpy and Heat Transfer Coefficient Basis for Example

Incremental Economic Heat Flux Basis

$$IHF = \frac{\text{ITHEC x Capital Recovery Factor}}{\text{IVES x Operating Hours per year}}$$

Incremental Total Heat Exchanger Cost=	~ $100/square foot
Incremental Value of Energy saved =	$4.0/MBTU
Capital Recovery Factor =	.25
Hours /year =	8320
IHF =	750
IHF at higher energy costs and lower UA costs =	100

This model can be used to estimate cost and value of steam based on optimum heat recovery economics rather than mechanical work economics. It was first built using MathCAD 2001™

for the heat transfer models and then transferred to Excel™ for the optimization. It was easier to make the model with MathCAD but easier to analyze the results with Excel.

Model Results

Changing the economic Incremental Heat Flux over the range from 750 (high exchanger cost--low fuel cost) to 100 (low exchanger cost—high fuel cost) decreases the furnace duty to about 60% of the base while the total surface area in the heat recovery zone increases by a factor of about 3.2 as shown in Figure 3.

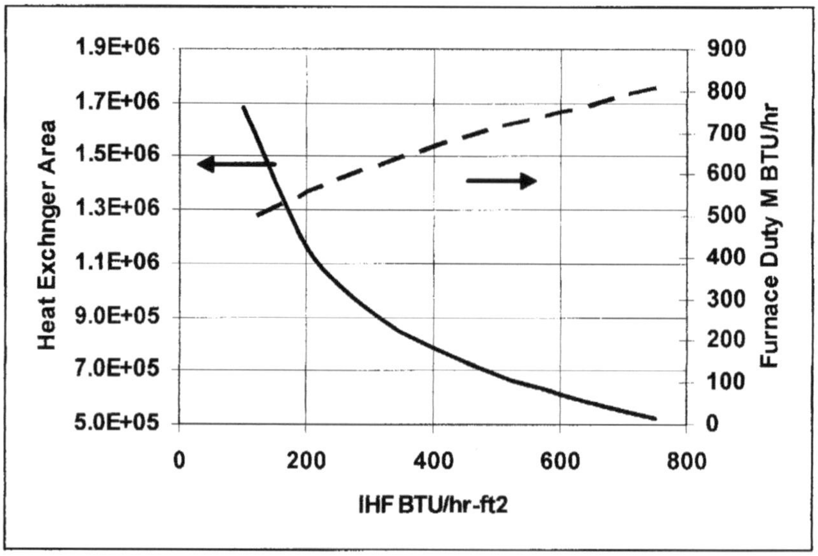

Generating up to 200 MBTU/hr steam in every case over the range of economic parameters studied caused an almost constant BTU of fuel increase for BTU of steam raised. Thus when steam is generated using the high temperature "waste heat" the cost of steam is essentially fuel cost and independent of the economic parameters. For heat transfer it is convenient to think about steam in heat content and temperature rather than pressure.

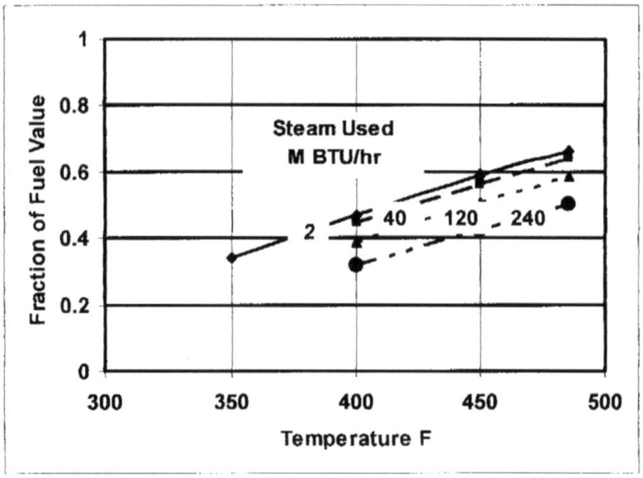

Figure 4 IHF =750

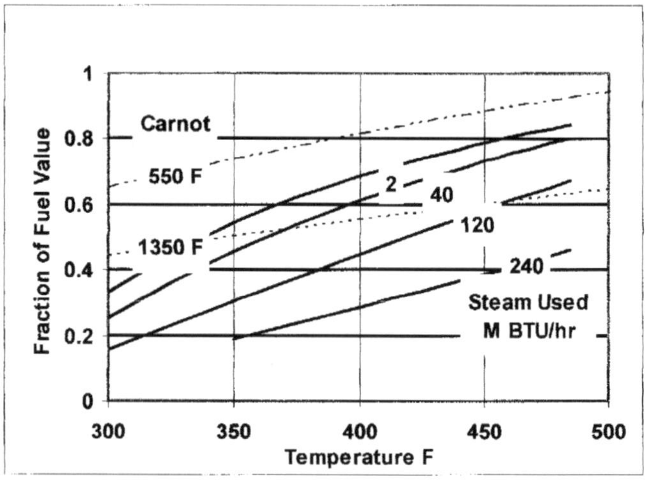

Figure 5 IHF = 100

Higher heat exchanger costs relative to fuel cost as represented an IHF of 750 (Figure 4) result in lower value of steam compared to IHF of 100 (Figure 5). The value of steam as a fraction of fuel value falls with temperature. For comparison the value of heat to a Carnot Engine with source temperatures of 550 and 1350 F and a sink temperature of 60 F is shown on Figure 5. The value of steam to heat recovery as calculated by the IHF method decreases more rapidly with temperature than the value calculated for a Carnot Engine. For this example the value of

300F steam is less than 30% of the value of fuel. It should find limited applications in heat recovery situations.

Figure 6 shows the value of 450 F steam as a function of the amount of steam used in the heat recovery zone. At 200 M BTU/hr input the value has decreased to about 45% of fuel value independent of the economic parameters. One can conclude that for this system the cost of cogenerated steam would need to be less than about 45 % of the cost of fuel for steam use to displace direct heat integration as the favored technology.

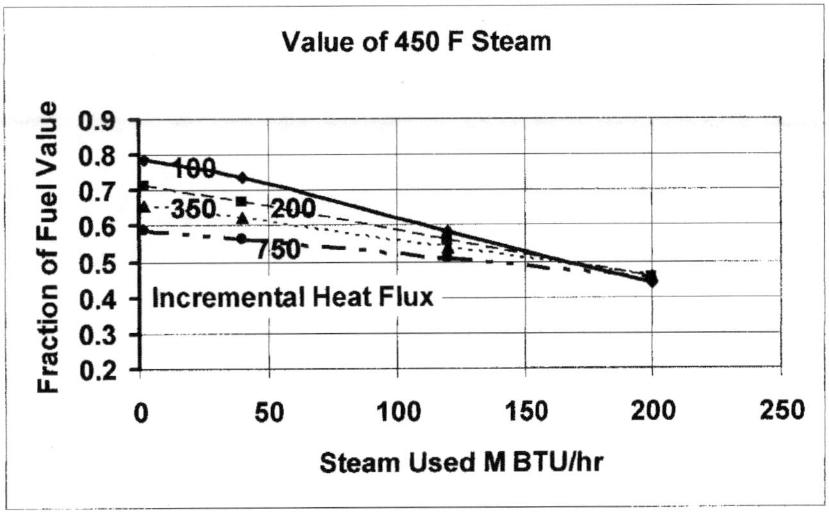

Conclusions:

An alternative method of estimating the value of heat from steam demonstrates that the value of heat to a process can be decidedly different from the cost calculated by thermodynamic methods. The alternative method uses the process characteristics in terms of composite curves, and heat transfer coefficients, along with the cost of heat exchangers and cost of fuel as a basis for estimating the value to the total system of replacing heat recovery from process by heat from an outside steam source.

---

[1] Cooper, D. Do you value steam correctly? *Hydrocarbon Processing* (July 1989)
[2] US DOE Industrial Technologies Program Best Practices Steam Technical Brief *How to Calculate the True Cost of Steam,* DOE/GO-102003-1736 (2003)
[3] Horlock, J. H. *Cogeneration—Combined Heat and Power (CHP), Thermodynamics and Economics,* Krieger Publishing Company, Malabar, FL (1997)
[4] Ghamarian, A; Thomas, W.R.L.; Sideropoulos, T; Robertson, J.L. Incremental Heat Flux Method for Heat Exchanger Optimization *Analysis of Energy Systems—Design and Operation* ASME Winter Meeting AES Vol. 1 (1985)

BOILING AND CONDENSATION IN PLATE HEAT EXCHANGERS

C.B. Panchal

Extended Abstract

Plate heat exchangers are being increasingly used for industrial applications as technical barriers are being removed and design codes are being developed. Some types of plate heat exchangers have been commonly used for specific industrial applications, such as brazed aluminum plate-fin heat exchangers in the cryogenic industry and plate frame heat exchangers in the dairy industry. Their applications are now widening in process industries. Plate heat exchangers are also known as compact heat exchangers because the area density is generally greater than that for shell-and-tube heat exchangers. Plate heat exchangers as condensers have volumetric coefficient of > 300 kW/m^{3}K as compared to < 200 kW/m^{3}K for shell-and-tube condensers. Table 1 shows comparison of shell-and-tube, plate-frame, and plate-fin heat exchangers for condensation of ammonia at comparable test conditions.

Table 1. Comparison of the condenser performance of three condensers

Heat Exchanger	Heat-transfer coefficient (kW/m^2 K)	Coolant-side pressure drop (kPa)	Area density (m^2/m^3)	Volumetric heat transfer coefficient (kW/m^3 K)
Shell-and-tube	2.7	20	71	192
Plate-fin	5.7[a]	21	69[b] 440[c] 102[d]	393
Plate-frame	4.2	103	191	802

a based on projected area b projected are
c condensing side d coolant side

Design flexibility for selecting geometric configurations and lower costs than shell-and-tube heat exchangers are primary factors for their increasing applications. However, greater design flexibility also imposes challenges for optimizing heat exchanger geometries for specific applications. Inadequate knowledge of two-phase flows in narrow passages of plate heat exchangers and fouling control further impose major challenges for designing plate heat exchangers for phase-change applications. This study focuses on salient features of two-phase flows in plate-frame and plate-fin heat exchangers and their fouling characteristics.

Two-phase patterns in narrow non-circular flow passages, which are typical of plate heat exchangers, cannot be predicted with high reliability using correlations developed for tubular geometries. Certain flow regimes may or may not exist in plate heat exchangers. The significance of two-phase flow pattern in plate heat exchangers are characterized by the three design considerations, 1) flow instability, 2) drainage of condensate without flooding, and 3) desired vapor/liquid flows for maintaining thermodynamic equilibrium for multi-component condensation and boiling.

The two-phase flow in plate-frame heat exchangers is difficult to predict; however, wavy thin film flow for down-flow condensation is a dominant flow pattern. The major concern for plate-frame heat exchanger is possibility of liquid flooding in the lower part. However, the performance tests of prototype units showed that condensate drainage is not a problem. In general, the outlet port should not be flooded with condensate for avoiding two-phase flow instabilities. The two-phase rising-film flow induced by flow boiling in plate-frame heat exchangers provides stable performance over a wide range of exist mass qualities. The outlet port should be adequately designed for two-phase outlet and/or separation.

The two-phase flow in plate-fin heat exchangers should not be a major concern. The fin density can be designed for different liquid and vapor flow rates. The flow visualization identified three main flow patterns 1) slug/reversing slug, 2) churn/transition, and 3) wavy film flows.

Fouling is a complex process that encompasses formation and transportation of foulant to the wall surface, and attachment and growth of the fouling film. Understanding of fouling characteristics of complex flow passages of plate heat exchangers is in its infancy. Complex flow passages are perceived to have high fouling propensity. However, lower fouling propensity has been observed for plate-frame heat exchangers than for shell-and-tube heat exchangers.

Paper 41e

SHELL-SIDE LONGITUDINAL VS. CROSS-FLOW

Blazo Ljubicic
Brown Fintube Company, EMEA
B.P: 83 / Zone Industrielle Riedgen
L-3401 Dudelange, Luxembourg
ljubicib@kochind.com

Key Words: shell-and-tube heat exchanger, shell-side flow

Prepared for Presentation at the 2004 National Meeting, New Orleans, LA, April 25-29, 2004

Unpublished

Shell-side Longitudinal vs. Cross-flow

Strengths and Weaknesses of Twisted Tube and Segmental Baffle Technologies

Abstract:

There are two main types of flows in shell-and-tube heat exchangers: cross-flow and longitudinal-flow. Though fundamentally different, the simple cross and longitudinal-flows have one common characteristic: to be efficient in industrial applications these two basic types of flows need constant redistribution of viscous and inertial forces. In order to maintain flow perpendicular to a bundle with multiple shell-side cross-passes, cross-flow needs constant direction change. Similarly, in order to be efficient, longitudinal flow needs a secondary flow component. It is precisely because of these phenomena that different ratios between the heat transfer rates and pressure losses are encountered in these two types of flow. Ultimately, the development of rationally optimized heat exchanger designs with new superior surface characteristics could be achieved only if the basic phenomena are well known and understood. In spite of considerable activity, the field of enhanced heat transfer is not "routine". In fact it is anything but ordinary. Experimentation is difficult, modeling of the phenomena is not very simple, and the most sophisticated analytical and numerical techniques are required to get results that are reasonably useful to practicing engineers. This paper discusses issues directly related to the structure of the flow and to the generation, development, and dissipation of turbulent perturbations and their significance in the design of baffled and twisted tube exchangers.

Introduction

The design of a heat exchanger requires controlling heat transfer coefficients for both fluid streams on both tube-side and shell-side. In shell-and tube exchanger, tube side is augmented by various types of inserts, while shell-side is usually controlled by a choice of baffle geometry and increased by extended surfaces. It is particularly difficult to manage heat transfer process on the shell-side where baffled exchanger geometry seldom yield favorable heat transfer enhancement characteristics.

The need for more efficient, smaller size exchangers in all varieties has resulted in the development of many heat transfer surfaces that claim to be more efficient than conventional tubes. This makes a selection of surface configuration for a particular application by no means a simple task. Just as most issues are seldom black or white, so are most good solutions seldom unequivocal. Characteristic of techniques is the fact that each is typically suitable only for some applications; in other words, a universally effective augmentation technique is nonexistent. There is no solution that makes one technology absolutely superior, but there are technologies that for particular applications are better than others.

Cross-flow exchangers

While having a good record of reliability and functionality, most conventional TEMA exchangers rely on standardized baffled geometries. The consequence of existence of baffles provided to support tubes and direct the flow across the tube bundle on the shell-side results in some notable limitations. As the pressure drop is wasted on reversing the flow, the cross-flow results in a relatively high pressure drop per unit of transferred heat. The high turnaround momentum loss around a baffle proportionally increases leaks through the tube-to-baffle and shell-to-baffle clearances. Furthermore, the shell-to-baffle leakage bypasses the bundle taking no active part in heat transfer until it re-mixes with the cross flow at some point at the outlet. Also, the shell-side flow field has as many re-circulation zones and dead, hot or cold spots, as there are baffles. These are areas where corrosion and/or fouling are initiated and zones with low heat transfer coefficient.

The importance of shell-side flow field in a cross-flow exchanger was first recognized by Tinker [1]. He pointed out that shell-side flow mal-distribution is significant in all flow regimes and that the most extreme cases exist in a laminar flow. The shell-side cross-flow field is controlled by number of different possible flow paths in the exchanger, mostly due to different momentum changes, with different resistances in each path. To estimate the amount of flow in each parallel stream, simultaneously solving pressure drop equations, each in terms of flow resistance coefficients, was required. Development of these was accomplished by the two well-defined extensive research programs, one at the University of Delaware and the other at HTRI.

The flow filed has an impact on the resulting mean temperature difference (MTD) or thermal effectiveness (ε). The basic assumptions for the derivation of both of these parameters are perfect radial and no axial mixing of the shell-side stream. In cross-flow within baffle compartment, there is a considerable axial mixing and in addition to the classical F correction factor for non-countercurrent flow, MTD requires a second correction factor (δ).

Longitudinal Flow in Twisted Tube Bundles

An attractive feature of the twisted tube, Fig.1, is that swirl is not as in other techniques, produced by a device attached to the tube. As such, they do not require an extra attention during assembly, maintenance, inspection, or cleaning.

These devices consist of helically twisted, double radius oval tubes, welded by their round ends to tube-sheets. The tubes contact one another at their wider sides, six times over the length of one twist pitch, which makes the unit practically vibration free. The purely longitudinal shell-side flow in twisted tube bundles is rarely mentioned in the theory of heat exchangers design, despite its ability to provide considerable surface density, low-pressure drop, and high heat transfer coefficient.

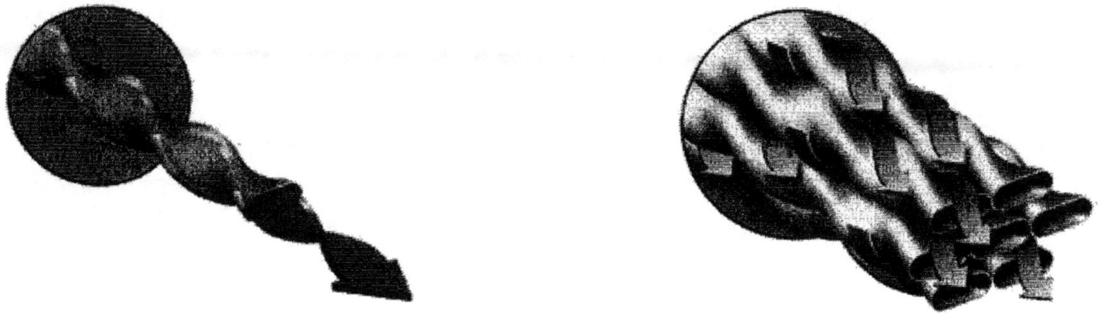

Fig.1 Tube-side and Shell-side Swirl Flow enhancement

The bundle enhances heat transfer, tube-side and shell-side. The tube-side fluid becomes swirled, and is affected by wall turbulence and by the different fluid layer velocities. In the course of the shell-side flow over tubes, circulation and mixing is generated. A secondary circulation generated by the centrifugal force of swirling also affects the in-bundle flow. Inside and outside helically shaped tubes double to triple the overall heat transfer rate with about doubled hydraulic resistance. A simple use of such tubes instead plane rounded tubes, is equivalent to a 30% reduction of the heat exchanger size while keeping a similar heat exchanger duty.

The action of active centrifugal forces promotes formation of random pulsation and increases the average flow turbulence, while in the zone of conservative forces the fluid viscosity damps most embryos of turbulence. As a consequence, if properly designed, a swirl flow device will have a geometry that produces active centrifugal force flow field. This will in turn promote the radial velocity change in a way that has positive effects on the heat and mass transfer processes. To allow optimization, a unique, twisted-tube swirl flow parameter, was defined.

Cross-Flow vs. Longitudinal Flow

Predictions about heat transfer and heat exchanger optimizations can be very complex, with layer of refinements and peculiarities, often buried in computer codes, and in some cases grossly misinterpreted. The high values of turbulent heat transfer coefficients can be traced to the generation of eddy, which in turn can be traced to the input of power; that is turbulent energy dissipation. Thus, it is not surprising that all turbulent heat transfer correlations are of the simple form, Eq. (1). These are empirical correlations which are based on the equation types derived from the analytical methods with constants and exponents adjusted to experimental data. This type of correlation is generally accepted for industrial design practice.

$$N_{u_x} = C \cdot \left(R_{e_x}\right)^m \cdot \left(P_{r_x}\right)^n \cdot \left(\Phi_x\right)^p$$

However, regardless of the correlation used, they all point out to the facts that the geometry with the smallest equivalent diameter always wins, and that uniform dissipation of energy always wins. This has been recognized in HTRI Report S-SS-3-4 which clearly states:

"The efficiency of producing heat transfer from available pressure drop depends on two major factors:

- The fraction of total energy expended in shear at the heat transfer surface
- The amount of recirculation or back-mixing that occurs to reduce the effective MTD at the HT surface

Therefore, shell-side geometry which minimizes re-circulating eddies, bypassing and leakage, produces the most effective use of pressure drop."

If the geometry takes an enormous pressure drop locally, such as in small windows in shell-and-tube exchangers, the results will suffer. If the geometry has relative dead zones due to bypassing, the results will suffer. If the geometry achieves length through number of tube-passes, the results will suffer because of the excessive loss of pressure at tube entries.

Having all these in mind, it is obvious that although complex, shell-side meandering flow in twisted tube bundles is predominantly axial with the pressure drop significantly lower than in cross-flow exchangers. A bundle is often shrouded to completely prevent any bypass flows and if needed to provide impingement protection. Compared to cross-flow exchangers, twisted tube longitudinal flow on the shell-side results in a higher heat transfer coefficients per unit pressure drop. Often heat transfer coefficients are 40% higher for the same pressure drop, or conversely, pressure drop is halved for the same heat transfer coefficient. The unidirectional, almost pure plug flow in twisted tube bundles results in designs with higher thermal effectiveness, typical of plate heat exchangers.

Conclusion

Although twisted tubes have been around for some time, new applications of twisted tube technology have been under development at Brown Fintube Company in Houston. Since 1994, Brown Fintube has been heavily committed to the development of this new and innovative technology and its heat transfer application in the process industry. With over 500 installations, in wide range of sizes and applications, it was clearly proved that improvement of the overall heat transfer rates was present for twisted tubes, as compared to available data and well established correlations for plain tubes.

The results to date indicate several advantages for this tubing concept:

1 The fabrication technique of simultaneously shaping and spiraling to form oval-twisted tubing results in relatively low fabrication costs, competitive to commercially available plain tubing.

2. The heat transfer coefficient is increased without an associated increase of the friction factor. The physical process is that of single-phase axial fluid flow. The twisted channels continuously induce rotation of the flow, both inside and outside the tube, thus enhancing turbulent momentum exchange in the immediate vicinity of the wall.

3. The friction coefficient does not increase proportionally with increased heat transfer. In fact it can be argued that at the same flow rate and for certain ratios of swirl to axial velocity friction factor for twisted could tube bundle could be lower than for a same size round tube bundle.

4. All tests show trend of increasing shell-side heat transfer coefficient. Favorable density gradient is an additional effect that will particularly benefit configuration with fluid condensing on the outside of the twisted tubes. In this case a twisted tube provides conditions which extends shear controlled regime over lower vapor fractions and channels for draining the condensate in gravity controlled regimes resulting in the major portion of the tube having a very thin condensate film. This in turn will reduce the resistance to heat flow and will substantially improve the heat transfer performance over conventional smooth tubes.

5. The combination of the enhancement on the outside of the twisted tube in condensation with the enhancement on the inside of the tube in boiling can result in the significant improvement in heat transfer coefficient. In this configuration the heat flow direction is radically inward, which leads to density gradient that are destabilizing, and as a consequence, turbulent exchange increases.

References:

1. Molloy, J., 1967, "Rough-tube Friction Factors and Heat Transfer Coefficients in Laminar and Transition Flow, " AERE-R 5415.
2. Round, G. F., 1980, Can. J. Chemical Eng., 58, Feb. pp. 122-123.
3. Simonds, R., 1968, "An Investigation of Heat Transfer in Smooth and Rough Tubes," Thesis in Mechanical Engineering, MIT.
4. White, A., 1964, "Flow of a fluid in an axially rotating pipe," J. Mech. Eng. Sci., Vol. 6, pp47-52.

VORTEX HEAT TRANSFER ENHANCEMENT FOR WASTE HEAT RECOVERY IN CHEMICAL INDUSTRY PROCESS HEATERS

Yaroslav Chudnovsky, Ph.D., Aleksandr Kozlov, D.Sc., Harry S. Kurek, P.E.
Gas Technology Institute, Des Plaines, IL, USA

Introduction

The overall objective of this work is to develop a competitive and cost-effective technology for significantly improving the energy efficiency of fired process heaters in Chemical Industry. Specific goals include a 3-5% increase in overall heater thermal efficiency with no significant increase in pressure drop or fouling rate. A number of approaches can be used to increase heat transfer rates in the convective section of fired process heaters. These approaches, in general, involve the incorporation of fins, baffles, turbulizers, etc. to increase either the heat transfer surface area or turbulence or both. These approaches are effective in increasing the heat transfer rates, however, this increase is invariably accompanied by an associated increase in convective section pressure drop as well as, for heaters firing 'dirty' fuels, increased fouling of the tubes - both of which are highly undesirable.

GTI has identified an approach, which will increase heat transfer rates without a significant increase in pressure drop or fouling rate. It involves the use of special 3D surface profile of heated tubes in the convection section (generally, spherical dimples pattern with certain arrangement). Compared to other types of heat transfer enhancement approaches, as illustrated in Figure 1, the proposed approach achieves very high heat transfer rates at the low pressure drops.

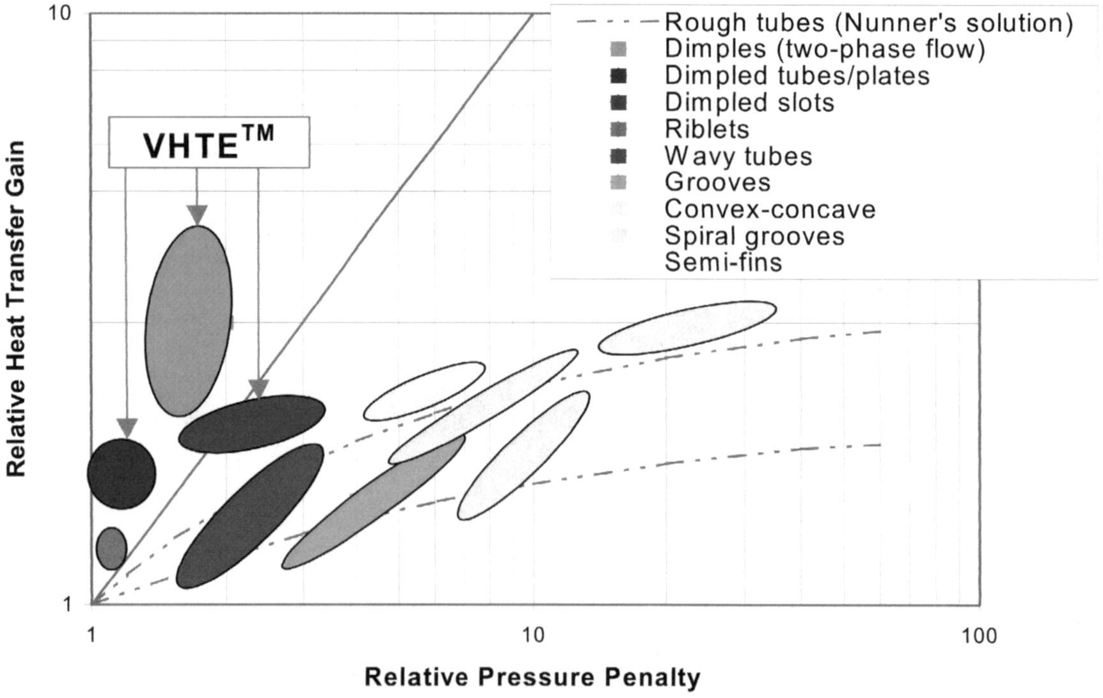

Figure 1. Efficiency of various heat transfer enhancement methods

Studies [1] have confirmed that while flowing a gaseous medium over surfaces with dimpled surface profile, each dimple works as a "vortex generator" that intensifies the rate of convective heat and mass transfer between the dimpled surface and the core of the gaseous medium (see Figure 2). The following benefits over bare surface were established for dimpled surfaces arranged in a tube bundle [2]:

- Increased heat transfer coefficient up to 20-30%
- No significant increase (or even reduction in some cases) in pressure drop

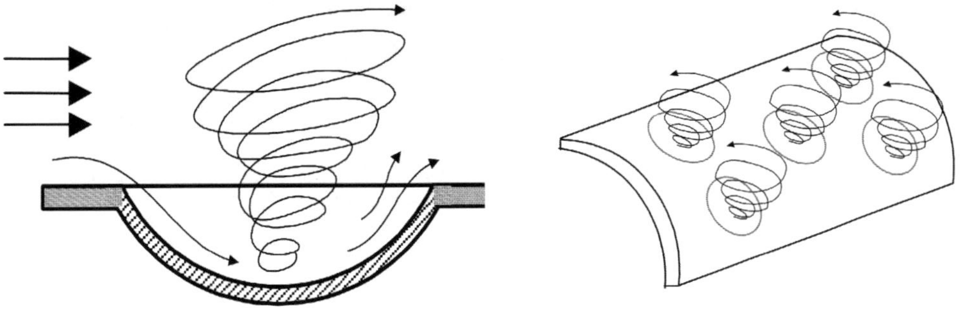

Figure 2. Mechanism of swirl formation in a spherical cavity and a cluster of vortices being developed on a dimpled tube surface [3]

Because of vortex generation inside the dimple the identified approach was named as Vortex Heat Transfer Enhancement (VHTE™).

Application Concept

Implementation of the VHTE™ technology into industrial applications will require the replacement of all the tubes in the existing (or new constructed) convective section with VHTE™ enhanced tubes (including shock tubes). The general layout of the originally proposed application concept is shown in Figure 3 below.

Development Status

The key hurdles of the development are:

- the risk of applying a new technology into an operating fired process heater;
- optimization of the dimple design to achieve maximum heat transfer rate increase at minimum or no pressure drop increase; and
- capital cost investment for energy savings.

During the continuing development three major steps were undertaken so far: CFD modeling to develop the efficient dimpled surface profile; experimental validation of the developed CFD model to prove the obtained results; and bench-scale evaluation following the comparative analysis of six test sections comprise of the different types of tubes: bare, finned and dimpled.

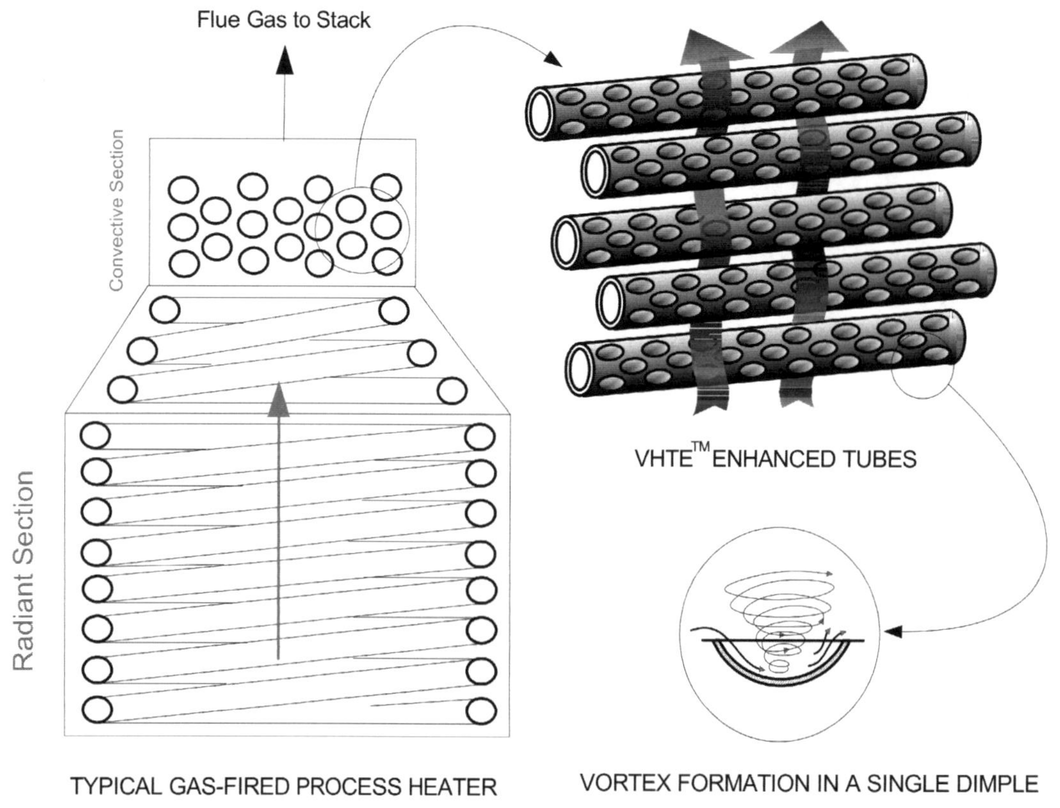

Figure 3. Application concept of VHTE™ enhancement of a fired heater convection section

CFD Modeling

The CFD modeling of hydrodynamics and heat transfer for a dimple tube bundle was performed by using a licensed copy of FLUENT 6.0 solver packaged with GAMBIT mesh generation code. GTI together with Fluent Inc. assistance developed and validated the computational model. The validation was based on the calculation results comparison with the available experimental data. Upon model validation GTI carried out extensive in-house parametric calculations for bare and dimpled tubes bundle at different flow conditions (inlet flow velocity and temperature) and dimple geometry (diameter and depth).

Based on GTI in-house CFD evaluation the RNG k-ε turbulence model with two layer zonal model was selected as better predicting heat transfer for bare and dimpled tubes in comparison with the k-ω turbulence model prediction.

Two different computational approaches (2D and 3D) were chosen to describe the tube banks computational domains. The 2D computational domain geometry was selected for bare tubes and a 3D modeling approach was selected for the dimpled tubes case. The details of the computational domains are shown in Figure 4.

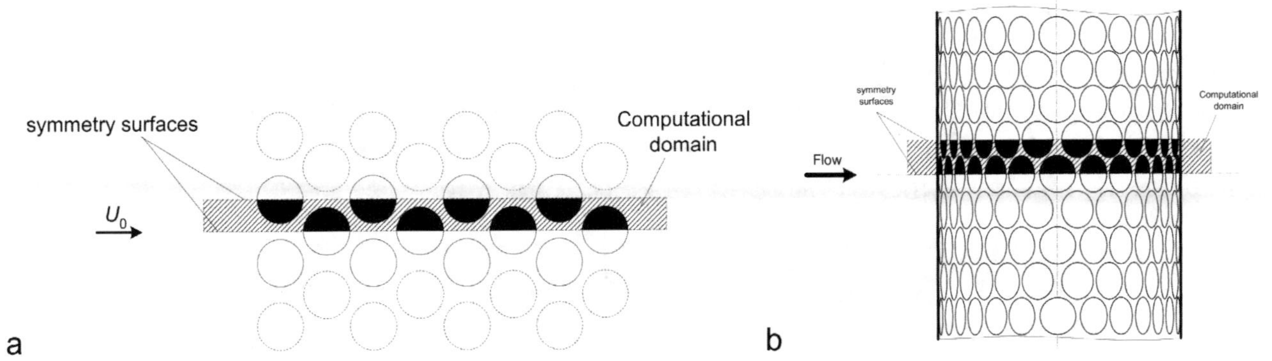

Figure 4. Computational domain geometry for bare (a) and dimpled (b) tubes

A uniform incoming velocity profile that is located at a distance of 5*D* from the first tube row was used at the tube bank inlet. Inlet turbulence intensity was ~1%. Inlet longitudinal and cross-flow turbulence integral scale was accepted as 3*D*. A uniform zero static pressure profile behind the tubes located at a distance of 10*D* from the last row was used at the outlet. Symmetry boundary conditions were used at the symmetry surfaces.

Figures 5 and 6 represent some modeling results on heat transfer and fluid flow over the dimpled tubes in a staggered arrangement.

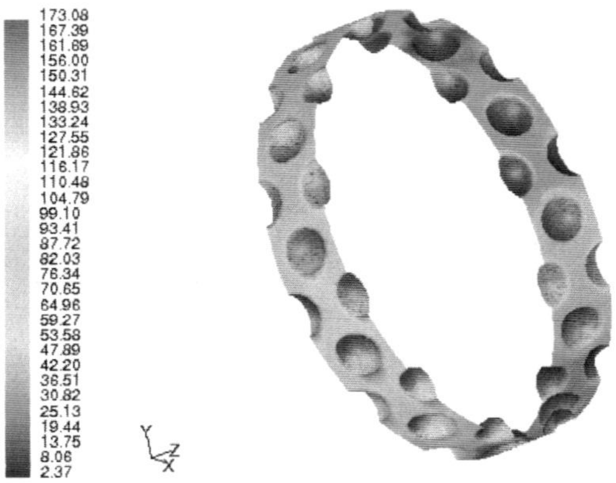

Figure 5. Heat transfer coefficient along the dimpled tube surface (slice view)

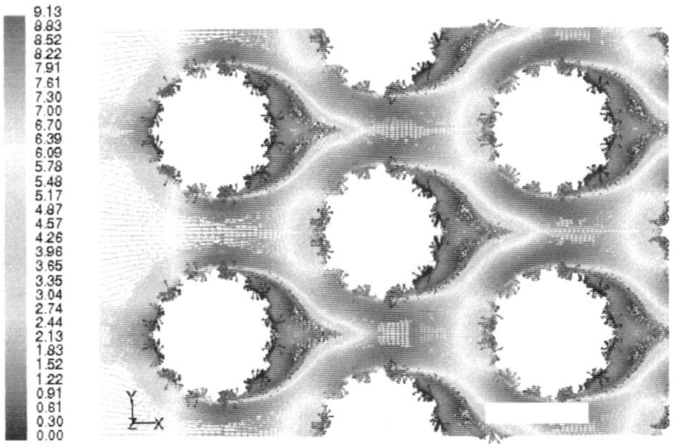

Figure 6. Velocity distribution between the dimpled tubes in a staggered bank

The essential heat transfer enhancement behind the dimple can be explained by the influence of the vortex, which is formed inside each dimple and escapes from the dimple. Those vortices increase turbulence intensity substantially and therefore positively affect heat transfer.

The modeling results for bare tubes were successfully validated with available reference data. The modeling results for dimpled tubes were acceptably validated with experimental results obtained in Convective Heat Transfer laboratory of the University of Utah.

Bench-Scale Unit

To evaluate bare, finned and dimpled tube banks in terms of heat transfer and pressure drop the Bench-Scale Unit (BSU) along with test sections (replaceable tube banks) were designed and fabricated by a local vendor. Figure 7 represents the general layout of the experimental rig installed at the GTI Applied Research Laboratory.

Figure 7. General layout of the experimental rig at the GTI laboratory

The entire experimental rig contained the following major components:

- air supply rig (0-25,000 SCFH)
- air heating system (0.6 MMBtu/h)
- water supply system (0-26 GPM)
- test sections (bare, finned and dimpled tube banks)
- measurement system (including the sensors and secondary devices)

A test section consists of a rectangular air channel with 20 tubes installed in a staggered arrangement. The pre-heated air (up to 1200°F) flowed over the tubes bank while city water flowed inside the tubes and provided heat removal from the BSU. Six test sections were developed and fabricated for BSU evaluation (see Figure 8 below representing three types of test sections).

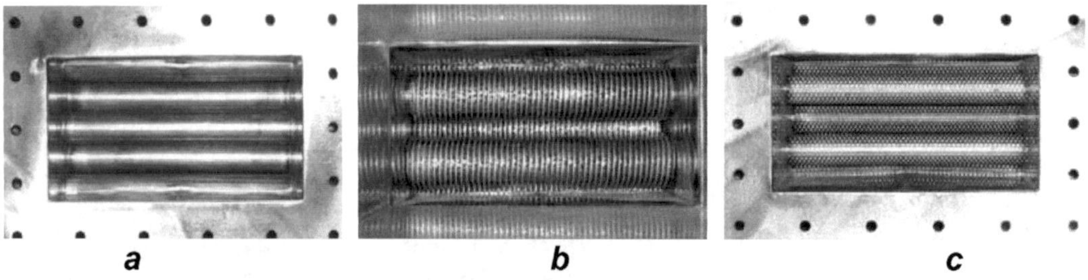

Figure 8. BSU test sections: **a** – bare, **b** – finned, **c** – dimpled

Each test section was equipped with pressure measurement taps, thermocouple couplings and water connection nipples. In order to minimize the possible heat losses to the ambient environment the test section was heavily insulated during the testing (see Figure 9).

Figure 9. Test section installed in BSU at the GTI laboratory

All the measurement sensors had 4-20mA outputs compatible with the Data Acquisition System (DAS). The measurement system included the following means for measuring of test parameters:

- thermocouples for measurement of air and water temperature
- flow meter for measurement of incoming air flow
- flow meter for measurement of water flow
- pressure transducers for pressure drop measurement across the test section

Measured signals were displayed on the DAS monitor along with the simultaneous recording to the data-files that were further converted into MS Excel spreadsheets for processing and analysis.

Comparative evaluations of all test sections were performed according to the following test approach. Incoming air was pre-heated by electrical heating system to about 1200°F and delivered to the BSU test section for heating water-cooled tube bank. The air flow rate was varied in order to evaluate heat transfer and pressure drop, at an incoming flow velocity in the range of 10-80 ft/s. City water flow rate through the test tubes was varied to keep water temperature below 150°F to prevent accidental formation of two-phase conditions.

During the test run the following parameters were measured, collected by the DAS and electronically recorded in output file (for each test section) for further processing and analysis:

- air flow rate, SCFH
- water flow rate, GPM
- air temperature before test section, °F
- air temperature after test section, °F
- water inlet temperature, °F
- water outlet temperature, °F
- pressure drop across the tube bank, in.WC

Major Test Results

The overall heat transfer coefficient and pressure drop across the tube bank for each of the six test sections were obtained. The results for dimpled tubes were compared with bare and finned tubes.

Heat Transfer

Based on temperature and flow measurements the following results were obtained for six test sections. Criterion Nu per bare tube diameter versus criterion Re constructed per inlet flow velocity are given in Figure 10. It is obvious that in the Re number range of 2500-4500 the heat transfer coefficient is 50-100% higher for bare tubes and 2-3 times higher for dimpled tubes as opposed to that of finned tubes.

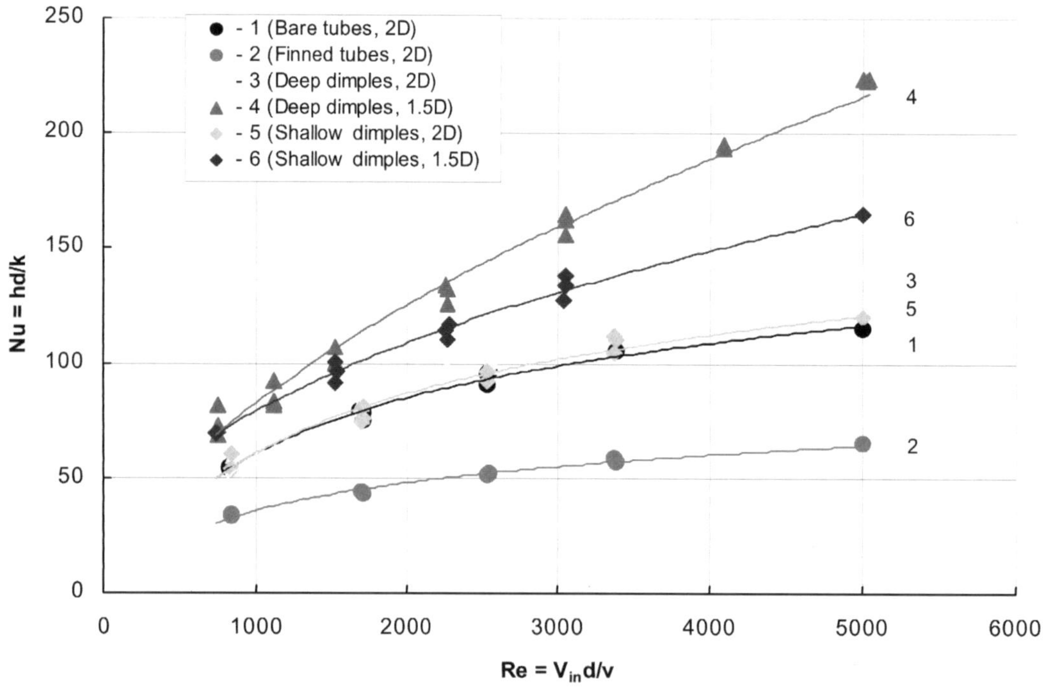

Figure 10. Heat transfer coefficient (Nusselt number) versus air velocity (Reynolds number)

Pressure Drop

Based on measurements of pressure drop across the test section the criterion Eu was constructed versus criterion Re that represents the inlet velocity measurement. Figure 11 demonstrates the obvious advantage of the dimpled tubes (for all dimple depths and profile densities) over the finned surface. In Re_{max} number range of 15,000-25,000 the pressure drop can be reduced up to 3 times by using the dimpled surface.

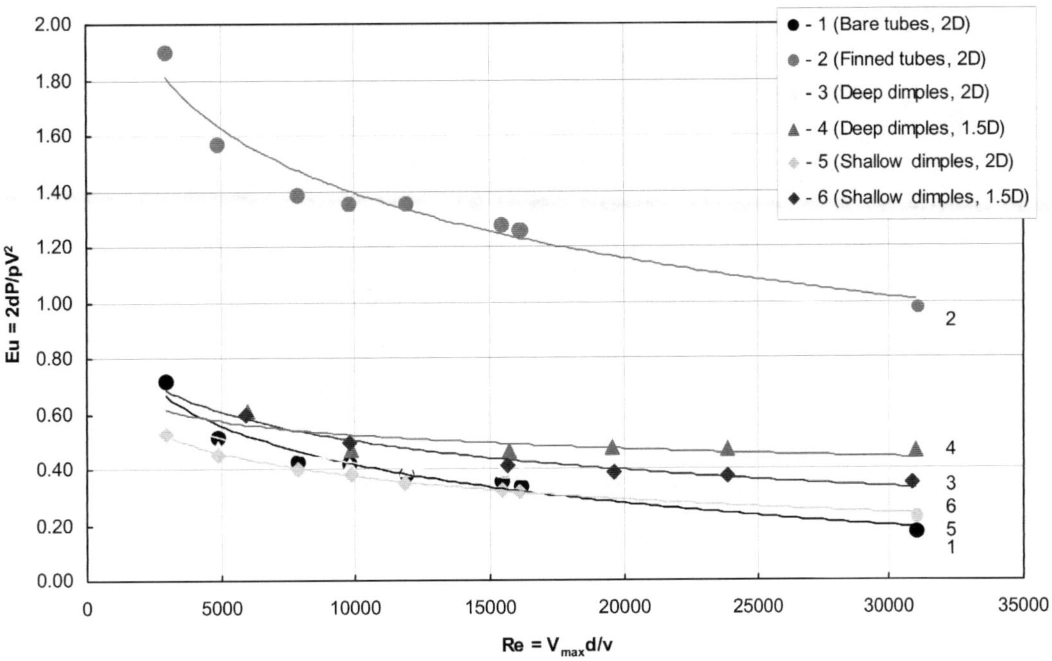

Figure 11. Pressure drop (Euler number) across the test section versus air velocity (Reynolds number)

Conclusions

The preliminary analysis demonstrates the strong potential for significant savings in energy (fuel), capital costs and increase in throughput by manufacturing or retrofitting the convective section for fired process heater with dimpled tubes as opposed to finned and shock tubes. Based on the experimental data the following benefits are estimated for the industrial application based on a natural gas price of \$5/MMBtu and a 700 high temperature fired heaters domestic population:

- energy savings per each 1% of fuel use decrease = \$250-350K per heater annually
- reduced expenses per each 1% retrofit of installed base = \$50-100K annually
- gain per each 1% increase in throughput = 10,000 tons/heater of product (assuming \$500/ton of ethylene ~ \$5M)
- lower operating costs due to reduced maintenance, cleaning, easy replacement, etc.

Currently GTI along with the Industrial Partners and Vendors are preparing for the field trial at the selected Chemical Industry site.

Acknowledgments

The authors are grateful to US Department of Energy (award DE-FC36-01ID14089), Gas Research Institute and Members of GTI's Sustaining Membership program for their financial support to this work.

References

1. A.Schukin, A.Kozlov, Y.Chudnovsky, R.Agachev. "Intensification of Heat Exchange by Spherical Depressions. A Survey", Applied Energy: Journal of Fuel, Power and Heat Systems, vol. 36, no. 3, 1998.

2. I.Shrader, A.Daschchyan, and M.Gotovskii. "High-Intensity Tubular Air Heaters", Thermal Engineering, vol. 46, no. 9, 1999.

3. Ya. Chudnovsky "Heat Transfer and Fluid Flow in a Spherical Cavity," Paper presented at Second European Thermal Sciences and UIT National Heat Transfer Conference, Rome, Italy, May 1996.

MODELING AND SIMULATION OF FLOW IN MICROFLUIDIC MICROREACTOR ARRAY DEVICE

*O. Srivannavit[1], S. O-Charoen[1], Z. Hua[1], X. Zhou[2], E. Gulari[1]**

[1]Department of Chemical Engineering, University of Michigan, Ann Arbor, MI 48109, USA
[2]Atactic Technologies Inc, 2212B Bellefontaine, Houston, TX 77030, USA
* Corresponding author (gulari@engin.umich.edu)

ABSTRACT

Modeling and simulation have been widely used in the design and optimization of complex microfluidic devices. In this study, an electrical equivalent network and a commercial CFD software package (CFD-ACE+ from CFDRC) are used to model and simulate a microfluidic microreactor array device. Geometrical variables have been studied including the microreactor depth and microchannel depth. The flow uniformity is tremendously improved by decreasing the microreactor depth, increasing the microchannel depth and/or changing the microchannel structure. In conclusion, by proper design of this microfluidic system, uniform flow and reaction in all microreactor cells can be achieved.

1. INTRODUCTION

Microfluidic devices are a rapidly emerging technology. Many components of these devices such as microchannels, diffusers, valves, pumps and flow meters [1] have been developed. Such devices have been used in a variety of applications that include biomedical dosing, biochemical reaction systems, and microfluidic mixing and regulation [2]. These miniaturized devices improve performance with respect to sample size, response time, and reagent consumption. There has been a growing interest in modeling and simulation of such devices because it can provide useful information for design of devices. In addition, this information can use to predict the performance of the device, to explain the experimental results and to optimize the design [3-8].

We designed, fabricated, and tested a microfluidic microreactor array device. This device was shown to be suitable for liquid phase synthesis of oligonucleotide and peptide microarrays [9]. In this paper, fluid flow in this device is modeled by an electrical equivalent network and a commercial CFD software package, namely CFD-ACE+ from CFD Research Corp, under steady state and incompressible flow. Several geometrical variables have been studied including the microreactor depth, microchannel depth and the shape of microchannels to develop a more profound understanding of the flow distribution through microscale reactors. The model was validated qualitatively with the evidence found in experimental results. In addition, we used the model to optimize shape of the microchannels to achieve more uniform flow distribution.

2. MODELING AND SIMULATION METHOD

To be able to model the microfluidic microreactor array, we need to understand the physical phenomena occurring in the device and identify simulation domains. This device is an external pressure driven device. As shown in Figure 1a, a fluid stream flows through the device by means of inlet microchannels and splits into side streams that enter microreactors along the inlet fluid microchannel. After passing through the microreactors, the side streams merge into the outlet fluid microchannel and flow out of the device. Because there are surface reactions during the synthesis in the individual microreactors, it is crucial to supply quantities of reagents uniformly to all microreactors in the device. In order to meet this criterion, we need to understand the relative flow rates and/or pressure drops across microreactors. Unfortunately, installation of on-chip micro-flow sensors is impractical since our device contains tens of microchannels and thousands of microreactors.

2.1. Electrical Equivalent Networks

For the steady state, incompressible fluid flow in pipe, the head loss (Δh) for length, L, of pine can be evaluated as

$$\Delta h = \frac{2\mu LP^2}{\rho g A^3} Q = kQ \qquad (1)$$

where, Δh is the head loss, Q is the flow rate of the fluid, P is the wetted perimeter,
A is the cross section area of the pipe, μ is the viscosity of the fluid,
ρ is the density of the fluid, and k is the fluid flow resistance

This equation can be analogous to an electrical resistance defined as

$$V = RI \qquad (2)$$

Therefore, we can use the electrical circuit network to represent the fluid flow in the device. Figure 1b represents a portion of our microfluidic microreactor array device as an electrical equivalent network shown in Figure 1a. Continuity equation at point A can be expressed as

$$Q_1 + Q_2 = Q_3 \qquad (3)$$

or

$$\frac{h1 - h2}{R1} + \frac{h3 - h2}{R2} = \frac{h2 - h4}{R3} \qquad (4)$$

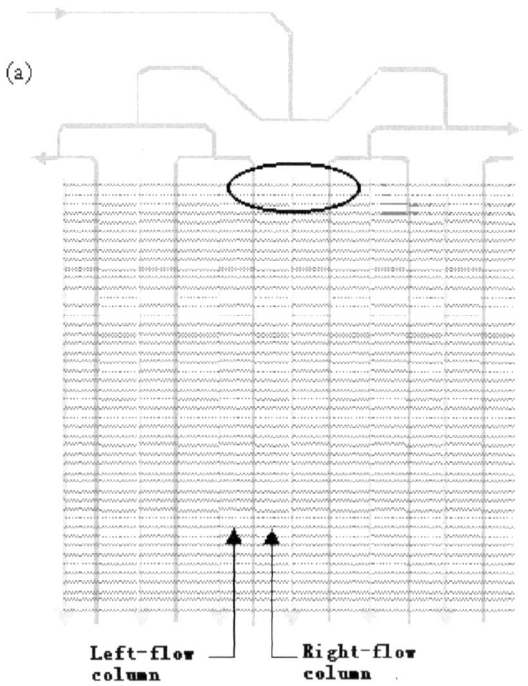

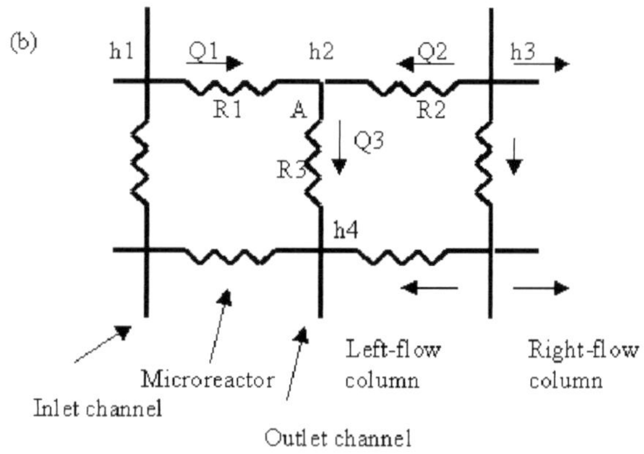

Figure 1. *(a) portions of the microfluidic array device; (b) electrical equivalent network of the microfluidic microreactor array device.*

We can obtain one equation at each node of the electrical network and a set of equations representing the device. A Matlab code was written to solve the set of equations simultaneously. The simulation output is the head (h) at each node of the electrical network. We can use this value to calculate the flow rate through the channels or microreactors using Equation 4.

2.2. Commercial CFD software package (CFD-ACE+)

This software numerically solves the transport equations for the 3-dimensional complex system. Figure 2 shows the model geometry of the microfluidic microreactor array containing 105 microreactors, which are divided into 7 columns and connected with 8 microchannels. The probe lines on the middle plane and at the center of microreactors in each column were used to compare the flow patterns of various geometrical variables. Figure 2b shows one of the microreactors with geometrical variables of interest.

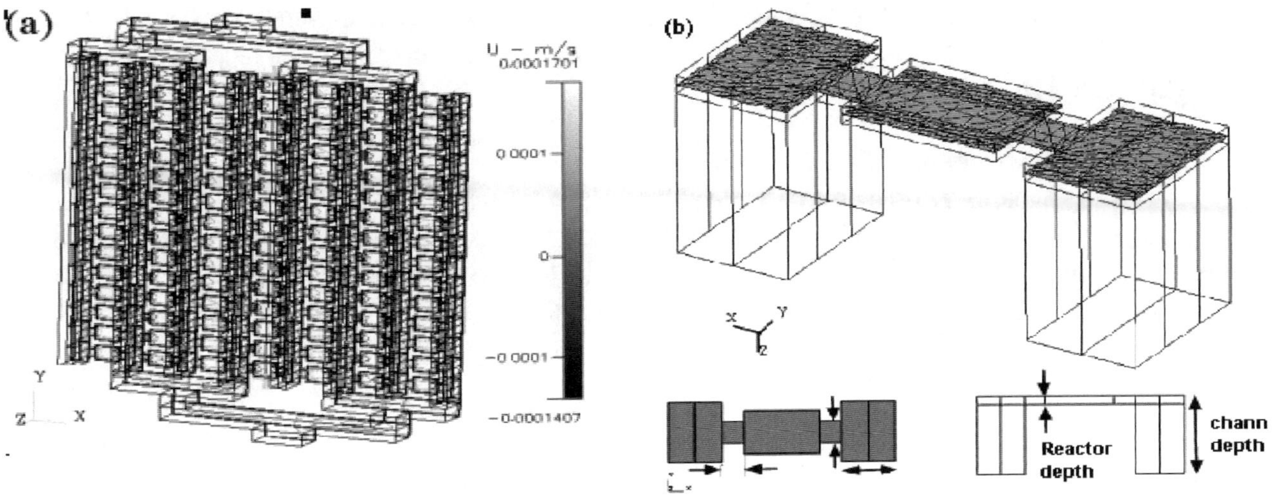

Figure 2. Model Geometry of the microfluidic reactor array generated from CFD-ACE+ software (a) 3-D geometry. (b) Individual microreactor with geometrical variables of interest.

3. EXPERIMENTAL METHOD

It is critical to validate the models by comparing the experimental results to model predictions. To reach this goal we used several different approaches as discussed below.

3.1. Fluorescent tracer experiment

In this experiment, the microfluidic device was mounted onto the real time chip adapter. The fluorescent dye source was connected to the adapter. Then, the fluorescent dye was pumped into the chip, which was initially filled with isopropyl alcohol. The chip adapter was put under an imaging camera to take images as a function of time. Figure 3a is the fluorescent image after three pumping cycles (each cycle carrying ~ 0.5 μL of fluorescent dye

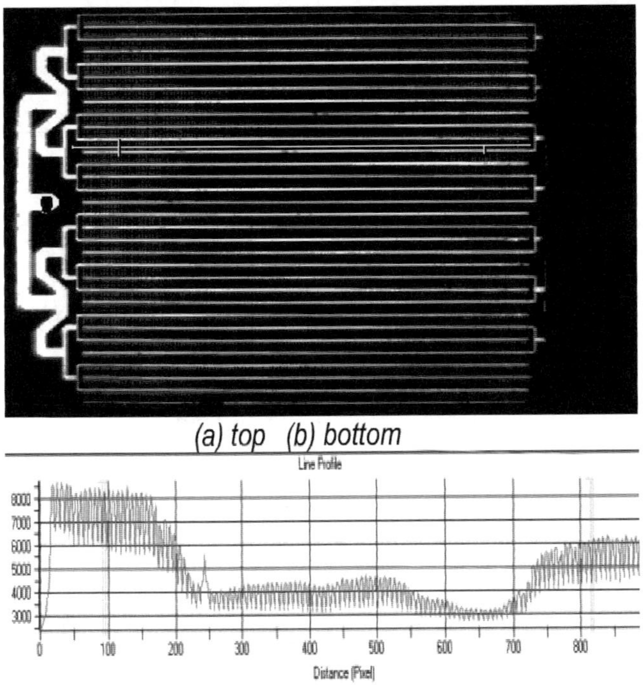

Figure 3. (a) Fluorescent image after three pumping cycles, 0.5 μl each, of fluorescent dye solution; (b) the line profile for the location specified in the fluorescent image.

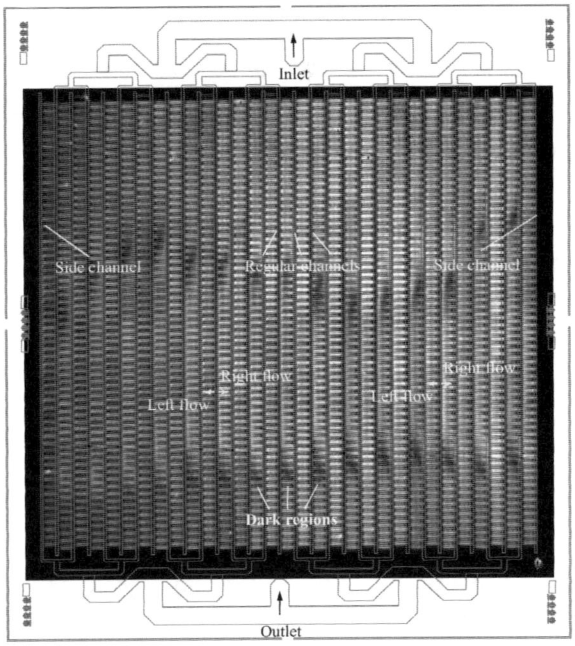

Figure 4. Fluorescent image of a chip containing fluorescein-tagged dimmer in a checkboard pattern

solution). Figure 3b is the line profile taken from one column of microreactors of the microfluidic device. As can be seen, the alcohol has been replaced by fluorescent dye more slowly in the middle of chip region (less fluorescent intensity) than in the inlet and outlet regions.

3.2. Validating evidence from the DNA synthesis chip

oDNA was synthesized in-situ on a microfluidic chip where the microreactor depth has been intentionally increased. Figure 4 shows a fluorescent image of a chip that contains fluorescein-tagged dimers arranged in a checkerboard pattern. In this chip, the bright spots are supposed to have the same intensity everywhere. However, we can see two dark regions in every other column. All the dark regions are in the right-flow columns whereas the left-flow regions are uniform. We believe that this results from a sharp decline in the flow of the reagents to that region during synthesis.

4. RESULTS AND DISCUSSIONS

Although the actual microfluidic device contains thousands of microreactors, the studied model is scaled to hundreds of microreactors to reduce computation time and resources. Matlab's code was used to simulate the flow pattern of the microfluidic device with eight inlet/outlet microchannels, each containing 64 microreactors. CFD-ACE+ was used to simulate the flow pattern of the device with eight inlet/outlet microchannels, each containing 15 microreactors. Figure 5 shows the flow rate pattern (microchannel depth = 100 μm, microreactor depth =10 μm) obtained from electrical equivalent network and commercial software simulation. The results show the non-uniform flow pattern observed experimentally. The flow is faster at the inlet and outlet ends of the chip and slower in the middle of the chip. In order to have high quality synthesis, the flow uniformity has to be improved. The parameters (i.e., microreactor depth and microchannel depth) which can be changed during fabrication were examined. We noticed that the flow uniformity can be degraded by increasing the depth of microreactor and/or decreasing the depth of microchannel. We also observed a disruption of flow distribution in the microreactors, especially in the right-flow column. As can be seen in Figure 6, the flow rate becomes zero at two locations of the right-flow column. In addition, the simulation results from the electrical equivalent network and the commercial CFD software indicate a change of flow direction from right to left. Based on simulation results, to avoid this non-uniform flow problem, we need to modify the dimesions of the device in such the way that the pressure drop across the microreactors is increased relative to the microchannels.

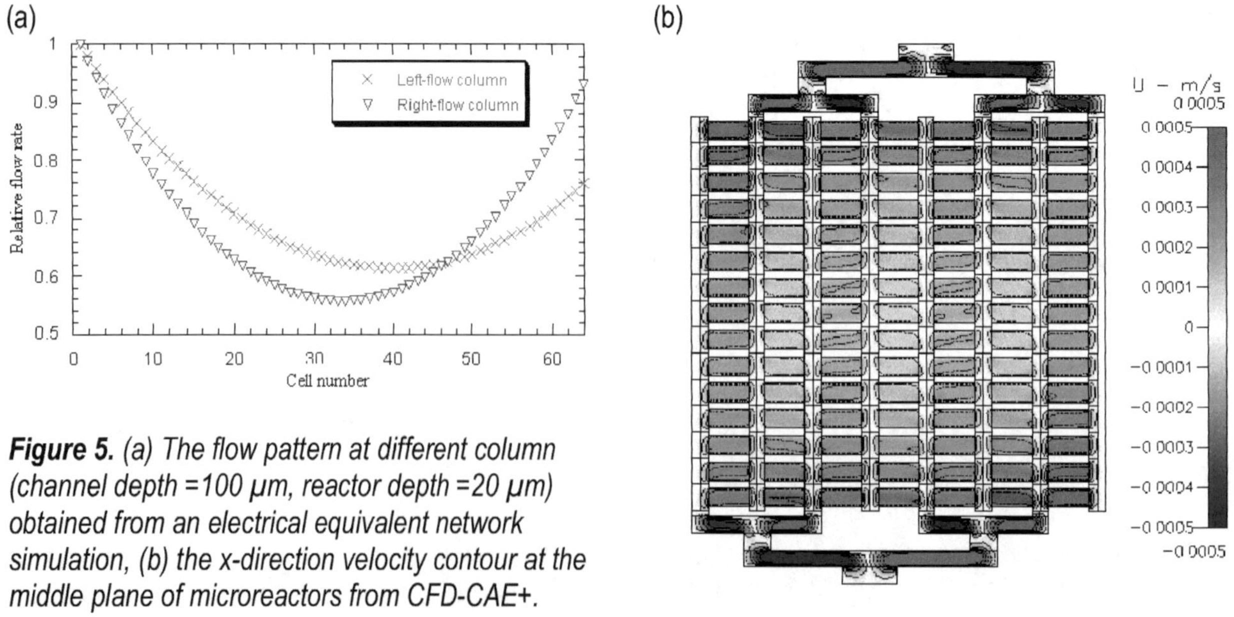

Figure 5. *(a) The flow pattern at different column (channel depth =100 μm, reactor depth =20 μm) obtained from an electrical equivalent network simulation, (b) the x-direction velocity contour at the middle plane of microreactors from CFD-CAE+.*

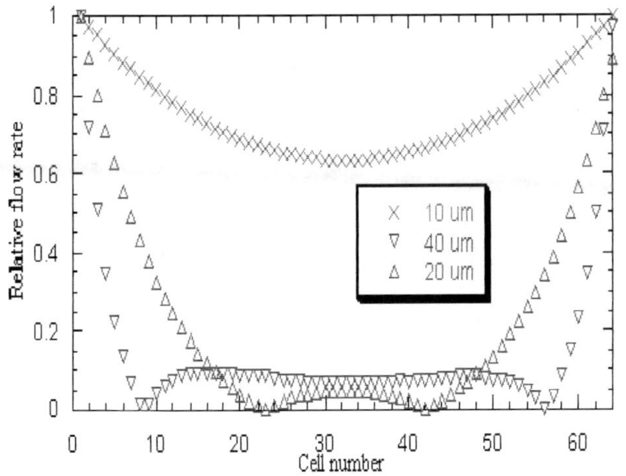

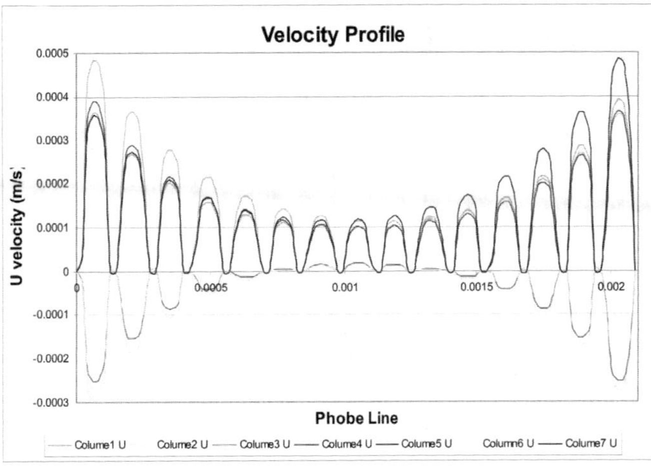

Figure 6. *(a) The flow pattern of the right flow column at different microreactor depths (channel depth = 100 μm) from an electrical equivalent network simulation (b) Plot of the x-direction velocity profiles along the probe lines at the center of reactor columns (left to right) in the backflow simulation case from commercial CFS-CAE+ (channel depth = 100 μm, reactor depth = 30 μm).*

5. OPTIMIZATION

Although our first generation of microfluidic microreactor devices could be used to manufacture oDNA chips, the level of uniformity of the reagents flow through the microreactors, evident in the simulation results indicate that the depth of microchannels and microreactors must be strictly controlled to achieve better uniformity in synthesis. We believe that the fundamental cause of the non-uniform flow pattern in the first generation of our device is a direct result of the way the liquid is distributed through the device. The side microchannels supply or drain liquid from only one side of the microchannels, whereas all regular microchannels symmetrically supply or drain liquid from both sides of the microchannels. In order to supply quantities of reagents to microreactors uniformly throughout the device, the device required either strictly controlled depth of the microchannels and the microreactors or a new design of microchannel shape, which is what we did. In this new design, fluid flow becomes unidirectional to microreactors for all the microchannels.

If we consider a device consisting of six microreactors with unidirectional fluid flow distribution, we can build the electrical equivalent network as shown in Figure 7(a). Since we do not know the profile of the microchannels, we assume that the unknown inlet microchannels have the same profile as the outlet microchannels, but in reverse, we will have six unknowns (a_1, a_2, a_3, a_4, a_5, a_6) in the electrical network. In order to solve the values of a_1-a_6, we need to have six equations.

Assuming that $h_0-h_1 = h_1-h_2 = h_2-h_3 = h_3-h_4 = h_4-h_5 = h_6-h_7$, we obtain the following equations:

$a_1(6Q) = a_2(5Q)$ or $a_2 = 6/5\, a_1$; $a_1(6Q) = a_3(4Q)$ or $a_3 = 6/4\, a_1$; $a_1(6Q) = a_4(3Q)$ or $a_4 = 6/3\, a_1$;
$a_1(6Q) = a_5(2Q)$ or $a_5 = 6/2\, a_1$; $a_1(6Q) = a_6(Q)$ or $a_6 = 6\, a_1$

Now, if a_1 is given, we can calculate the remaining of the values of a_2-a_6 using the above equations. Figure 7(b) shows the profile of microchannels with inlet microchannel end width of 100 μm. As can be seen in the same Figure, the profile is not significantly changed when the microchannel depth is changed. In other words, the flow distribution in this new design is less sensitive to the set depth of the channels.

We fabricated a microfluidic device with this new microchannel profile design. We performed a fluorescent tracer experiment on the device and the preliminary result indicates that there is uniform flow distribution through the device.

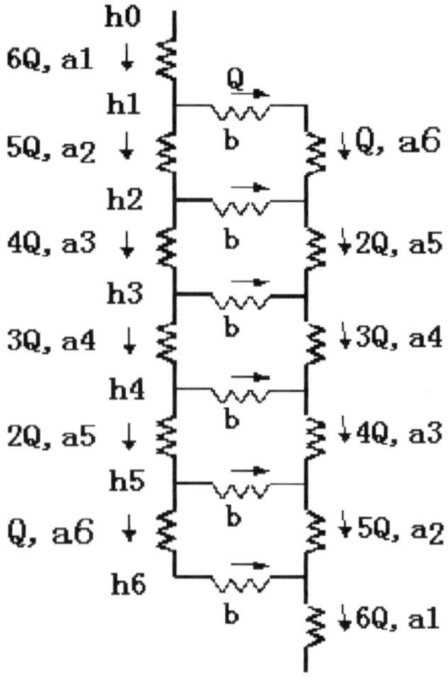

(a)

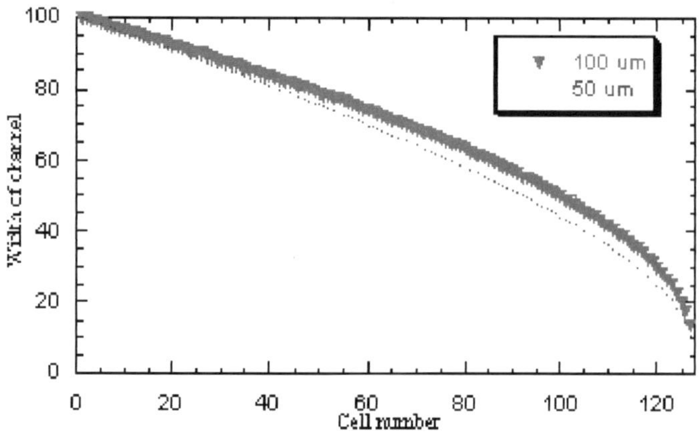

(b)

Figure 7: *(a) Electrical equivalent network for six microreactors with unidirectional flow distribution; (b) Tapered channel profile at different channel depth.*

6. CONCLUSIONS

Using a simple electrical equivalent circuit network and commercial software (CFD-ACE+), we were successful in modeling our microfluidic microreactor array device. We validated the model by comparing the simulation results with qualitative, experimental observations. We demonstrated that modeling can be applied to design new generation device for achieving the desired characteristics of uniform reagents distribution to microreactors throughout the chip.

ACKNOWLEDGEMENTS

Partial financial support of this research by a Michigan Life Sciences grant (to Erdogan Gulari) and an NIH SBIR grant (to Xiaochuan Zhou) is gratefully acknowledged.

REFERENCES

[1] A. Rasmussen, C. Mavriplis, M. E. Zaghloul, O. Mikulchenko, and K. Mayaram, "Simulation and optimization of a microfluidic flow sensor," *Sens. acutators A*, vol. 88, pp. 121-132, 2001.

[2] N. Vandelli, D. Wroblewski, M. Velonis, and T. Bifano, "Development of a MEMS microvalve array for fluid flow control," *J. Microelectromechanical systems*, vol. 7, pp. 395-403, 1998.

[3] M. Elwenspoek, T. S. J. Lammerink, R. Miyake, and J. H. J. Fluitman, "Towards integrated microliquid handling systems," *J. Micromech. Microeng.*, vol. 4, pp. 227-245, 1994.

[4] E. S. Piekos and K. S. Breuer, "Numerical modeling of micromachanical devices using the direct simulation Monte Carlo method," *Transactions of the ASME*, vol. 118, pp. 464-469, 1996.

[5] M. Carmona, S. Marco, J. Sieiro, O. Ruiz, J. M. Gomez-Cama, and J. Samitier, "Modelling of microsystems with analog hardware description languages," *Sens. acutators A*, vol. 76, pp. 32-42, 1999.

[6] N. R. Swart and A. Nathan, "Flow-rate microsensor modeling and optimization using SPICE," *Sens. acutators A*, vol. 34, pp. 109-122, 1992.

[7] H. A. C. Tilmans, "Equivalent circuit representation of electromechanical transducers: I. Lumped-parameter systems," *J. Micromech. Microeng.*, vol. 6, pp. 157-176, 1996.

[8] F. J. Auerbach, G. Meindres, R. Muller, and G. J. E. Scheller, "Simulation of the thermal behavior of themal flow sensors by equivalent electrical circuits," *Sens. acutators A*, vol. 41/42, pp. 275-278, 1994.

[9] O. Srivannavit, "Design, fabrication and modeling of microreactor arrays for biochips and discovery research", PhD Dissertation, University of Michigan, 2002.

An Overview of Biomass Gasification

Richard L. Bain,
National Renewable Energy Laboratory, Golden, Colorado, USA

Introduction

The basic principles of biomass gasification will be presented. Applications discussed will include heat and power generation, and synthesis gas production. As received, biomass can range from very clean wood chips at 50% moisture, to urban wood residues that are dry but contaminated with ferrous and other materials, to agricultural residues, to animal residues, sludges, and the organic component of municipal solid waste (MSW). Representative biomass compositions are compared to fossil fuel compositions in Table 1. The process of gasification can convert these materials into carbon- and hydrogen-rich fuel gases that can be more easily utilized, often with a gain in efficiency and environmental performance compared to direct combustion of the biomass.

Table 1: Representative Biomass & Fossil Fuel Compositions

	Poplar	Corn Stover	Chicken Litter	Black Liquor	IL. No 4 HvBb	Rosebud Sub B	Athabasca Bitumen
Proximate, wt% as received							
Ash	1.16	4.75	18.65	52.01	12.88	7.82	
Volatile Matter	81.99	75.96	58.21	35.26	37.54	33.32	
Fixed Carbon	13.05	13.23	11.53	6.11	42.04	41.91	
Moisture	4.8	6.06	11.61	9.61	7.54	16.94	
HHV, Dry (Btu/lb)	8382	7782	6310	4971	12,400	11,684	17,900
Ultimate, wt% as received							
Carbon	47.05	43.98	32.00	32.12	63.43	62.59	83.6
Hydrogen	5.71	5.39	5.48	2.85	5.10	6.27	10.3
Nitrogen	0.22	0.62	6.64	0.24	1.09	1.08	0.4
Sulfur	0.05	0.1	0.96	4.79	0.11	0.11	
Oxygen (by diff)	41.01	39.1	34.45	0.71	12.98	12.85	0.2
Chlorine	<0.01	0.25	1.14	0.07	4.40	4.36	5.5
Ash	1.16	4.75	19.33	51.91	12.88	12.75	
H/C Atomic Ratio	1.45	1.46	2.04	1.06	0.96	1.19	1.47
Elemental Ash Analysis, wt% of fuel as received							
Si	0.05	1.2	0.82	<0.01			
Fe	---	---	0.25	0.05			
Al	0.02	0.05	0.14	<0.01			
Na	0.02	0.01	0.77	8.65			
K	0.04	1.08	2.72	0.82			
Ca	0.39	0.29	2.79	0.05			
Mg	0.08	0.18	0.87	<0.01			
P	0.08	0.18	1.59	<0.01			
As (ppm)			14				

Gasifier systems incorporate a biomass fuel handling and feeding system, which is normally coupled by means of airlocks to the gasifier. The gasifier is usually a refractory lined vessel and fluid-bed or entrained-flow gasification is carried at temperatures of 750-850°C at either atmospheric or elevated pressures. The product gas has to be treated so it matches the end-use application. For close-coupled gasifier-combustor systems there is no cleanup of the gases. For gas turbine applications in a power system the gas has to be free of particulates, tars, sulfur, chlorine compounds, and alkali metals to ensure the integrity of the turbine hot section. For internal combustion engines the gas must be cooled to ensure that a sufficient charge of energy can be put into each cylinder; and particular attention has to be given to tar and particulate contents to ensure that the valves and cylinders are protected. Fuel cell applications require the gas to be mainly hydrogen without any significant sulfur or chloride contamination to protect the electrodes. For synthesis operations such as methanol and hydrogen production, particulates and other contaminants (H_2S, etc.) must be removed to prevent poisoning of downstream catalysts.

The process efficiency is high; gasification is typically 80-85% thermodynamically efficient in converting the organic content of the feed into a fuel gas mixture. Because biomass gasification results in a clean fuel gas, the efficiency is further enhanced by the use of combined cycles to generate electricity; biomass-to-electricity efficiencies greater than 45% are forecast. If the gases are converted to hydrogen, the limiting efficiency with fuel cells may be over 55%. The environmental advantage is that the fuel gas is a much smaller volume to be processed than the combustion stream from a boiler; this and the generally lower treatment temperature of the biomass results in retention of metals (including alkali) in the ash and cyclone as salts that can be disposed. The gas can easily be cleaned of acid gas components, including hydrogen chloride, before combustion and thus is environmentally superior to direct combustion.

Biomass is a complex mixture of organic compounds and polymers (Graboski and Bain 1979). The major types of compounds are lignin and carbohydrates (cellulose and hemicellulose) whose ratios and resulting properties are species dependent. Lignin, the cementing agent for cellulose, is a complex polymer of phenylpropane units. Cellulose is a polymer formed from d(+)-glucose; the hemicellulose polymer is based on hexose and pentose sugars. Biomass such as wood typically has low ash, nitrogen, and sulfur contents. However, some agricultural materials such as straws and grasses have substantially higher amounts. To estimate yields during gasification the complex material must be reduced to a simplified chemical formula, such as $CH_{1.4}O_{0.6}$. In this discussion, elements such as sulfur and nitrogen are considered to be present in very small amounts, and are not considered in terms of overall chemistry.

The combustion of biomass can be ideally represented by:

$$CH_{1.4}O_{0.6} + 1.05 * O_2 \rightarrow CO_2 + 0.7 * H_2O$$

Oxygen-blown gasification can be thought of as incomplete combustion or partial oxidation. Gasification using a minimum amount of oxygen can be represented by:

$$CH_{1.4}O_{0.6} + 0.2 * O_2 \rightarrow CO + 0.7 * H_2O$$

In cases where no oxygen is used the gasification reaction can be represented by:

$$CH_{1.4}O_{0.6} \rightarrow 0.6 * CO + 0.4 * C + 0.7 * H_2O$$

This pyrolysis reaction is endothermic, and heat is needed to make the reaction proceed. This heat is provided by the oxidation reactions shown above or by indirect heat transfer. Although these ideal reactions are simple, actual gasification is more complex. Intermediate compounds such as tars and methane are formed that must be further processed if the synthesis gas is be used to produce methanol or hydrogen.

Gasifier Medium

Gasification involves the thermal destruction of biomass in a reducing atmosphere of steam or air (or both) to produce a medium- or low-calorific value gas. If air is present, the ratio of oxygen to biomass is typically around 0.3. Air-blown, or directly heated gasifiers, use the exothermic reaction between oxygen and organics to provide the heat necessary to devolatilize biomass and to convert residual carbon-rich chars. For directly heated gasifiers, the heat to drive the process is generated inside within the gasifier. When air is used, the product gas is diluted with nitrogen and typically has a dry-basis calorific value of 5-6 MJ/Nm^3. The dry-basis calorific value of the product gas can be increased to 13-14 MJ/Nm^3 using oxygen instead of air. Oxygen production is expensive, however, and its use has been proposed only for applications involving the production of synthesis gas where nitrogen is not permitted in downstream synthesis conversion operations. Indirectly heated gasifiers heat and gasify biomass through heat transfer from a hot solid or through a heat transfer surface. Because air is not introduced into the gasifier, little nitrogen is present and a medium-calorific gas is produced; dry-basis values of 18-20 MJ/Nm^3 are typical.

Pressure

A second variable affecting gasification-based power systems performance is operating pressure. Most gasification power applications will involve a gas turbine-based generation cycle. Turbines typically operate at compression ratios of 10 to 20, giving turbine inlet pressures of 1.0-2.5 MPa. A pressurized gasifier will produce gas at a pressure suitable for direct turbine application and provide the highest overall process efficiency, but process improvements are needed. Reliable, high-pressure feed systems have not been commercially proven. At gasifier conditions (825°C and 2 MPa) tars, chars, and volatile alkalis are generated. To maximize efficiency, these materials must be removed from the hot product gas without lowering the temperature below the tar dew point, typically about 540°C. Thus, hot-gas cleanup systems are required. Tars have relatively high heat contents and can be burned in combustors, but they may plug char filters and may form soot during combustion. Therefore, the first element of a hot gas cleanup system will probably be a catalytic or thermal tar cracker. A catalytic tar cracker will operate at temperatures comparable to gasifier temperature, about 825°C; a thermal cracker will typically operate at 870-980°C. After the tar cracker, the product gas will be partially cooled to minimize the amount of alkali vapors, typically to 350-650°C. The product will then pass through a filter to remove solids. Much of the alkali will also be removed. For gas turbine applications the gas may still contain too much alkali. Gas turbine limits for alkali are about 25 ppb in the turbine combustor exit gas. To ensure that the product gas meets gas turbine manufacturer alkali specifications, an alkali getter bed may be added after the filter. Typical bed materials are emalthite or hectorite.

Alternatively, the gasifier can be operated at low pressure and the cleaned product gas compressed to the pressure required for gas turbine application. In this case a tar cracker will probably be used to minimize the amount of tar that must be handled during quenching. The product gas exiting the tar cracker will be conditioned to provide a suitable compressor feed. Gas characteristics to be considered for compressor use include the amount of condensable or soluble tar and the water vapor content. The water and tar content must be low enough to ensure no condensation during compression. Also, soluble tars that will affect lubrication oil properties must be eliminated. Lastly, the gas inlet temperature must be suitable for compressor materials of construction. Normally, the temperature should be held below about 90°C. Usually a combination of heat exchange (to reduce the gas from tar cracker exit temperature to residual tar dew point) and wet scrubbing are used. The water vapor content will be at saturation at scrubber exit temperature and pressure.

Gasifier Type

The other major variable is reactor type. Four primary types of biomass gasification reactor systems have been developed: fixed-bed reactors, bubbling fluid-bed reactors, circulating fluid-bed reactors, and entrained-flow reactors. Simple schematics of the various gasifiers are shown in Figure 1.

Fixed Bed

Fixed-bed gasifiers can be classified primarily as updraft and downdraft. Updraft gasifiers represent the oldest and simplest gasifiers. The updraft gasifier is a counterflow reactor in which fuel is introduced into the top by means of a lockhopper or rotary valve and flows downward through the reactor to a grate where ash is removed. The gasifying medium, air or oxygen and possibly steam, is introduced below the grate and flows upward through the reactor. At the bottom of the reactor (combustion zone) char burns to form carbon dioxide (CO_2) and steam (H_2O), which then flow upward through the bed countercurrently to the downflowing solids. The exothermic combustion reactions supply the energy to drive gasification, pyrolysis and drying. The maximum temperature in the combustion zone is typically higher than 1,200°C. In the reduction zone CO_2 and H_2O are partially reduced to carbon monoxide (CO) and hydrogen (H_2) through reaction with carbon in the char at temperatures of 800-1,200°C. In the pyrolysis zone these gases contact dry biomass in the temperature range of 400-800° C and devolatilize the biomass to produce pyrolysis products and residual char. Above this zone the gases and pyrolytic vapors dry the wet biomass. Typical product exit temperatures are 80-100°C. A wide range of tars and oils, which can condense in product lines, are produced in the pyrolysis zone. For this reason updraft gasifiers are usually operated in a close-coupled mode to a furnace or boiler to produce steam or hot water. Certain feeds with low-melting ash may have slagging on the combustion grate. In addition, feed particle size needs to be controlled to maintain a uniform bed.

Figure 1: Gasifier Types

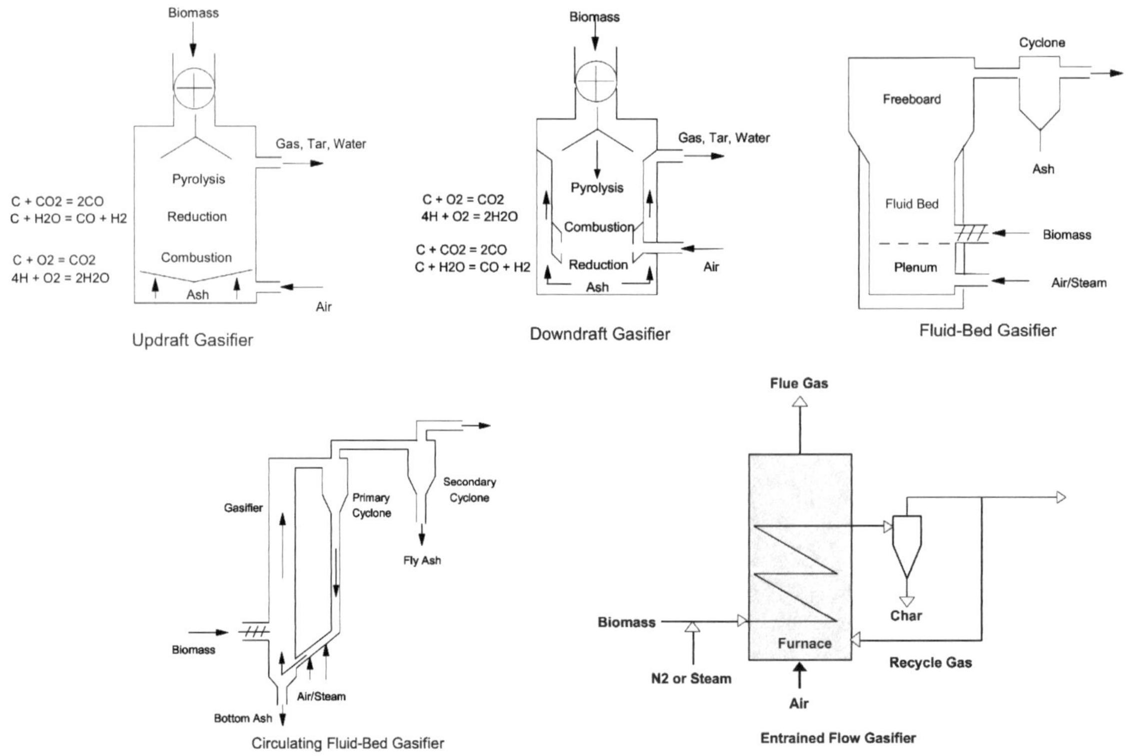

In downdraft gasification the air and product both flow in the same direction as the solid bed. Downdraft gasifiers are specifically designed to minimize tar and oil production. The fuel and pyrolytic gases/vapors move co-currently downward through the bed. The pyrolysis products pass through a hot char (about 15% of the original feed) combustion zone, where they are contacted with air and the tars are thermally cracked and partially oxidized. Typical tar conversion is greater than 99%, and is a function of temperature, combustion efficiency and channeling. The combustion zone temperature is typically 800-1,200°C. The hot char in the reduction zone reduces CO_2 and H_2O, to CO and H_2. The exit gas temperature is typically around 700°C. Downdraft gasifiers have the same general constraints on feed properties as updraft gasifiers. The feed needs to have a fairly uniform particle size distribution with few fines to maintain bed physical properties and minimize channeling. The feed needs to have low ash with a high fusion temperature to prevent slagging. In addition, the feed moisture content needs to be less than about 20% to maintain the high temperatures required for tar cracking. A variation on the downdraft gasifier is the crossflow gasifier, in which air is introduced tangentially at the bottom of the gasifier. The operating principle of the crossflow gasifier is the same as for the downdraft gasifier.

Bubbling Fluid Bed

In a gas-solid fluidized bed a stream of gas passes upward through a bed of free-flowing granular materials in which the gas velocity is strong enough that the solid particles are widely separated and circulate freely throughout the bed. During overall circulation of the bed, transient streams of gas flow upward in channels containing few solids, and clumps or masses of solids flow downward (Perry and Chilton 1973). The fluidized bed looks like a boiling liquid and has the physical properties of a fluid. In fluidized-bed gasification of biomass the gas is air, oxygen, or steam, and the bed is usually sand, limestone, dolomite, or alumina. The gas acts as the fluidizing medium, and if air/oxygen is used, is the oxidant for biomass partial oxidation. A fluidized-bed gasifier is a vessel with dimensions such that the superficial velocity of the gas maintains the bed in a fluidized condition at the bottom of the vessel, with a change in cross-sectional area above the bed that lowers the superficial gas velocity below fluidization velocity to maintain bed inventory and act as a disengaging zone. To obtain the total desired gas-phase residence time for complete devolatilization, the larger cross-sectional-area zone (usually referred to as the freeboard) is extended. A cyclone is used to either return fines to the bed or to remove ash-rich fines from the system. A gas distribution manifold or series of sparge tubes are used to fluidize the bed (Hansen 1992). Biomass is introduced either through a feed chute to the top of the bed or through an auger into the bed. In-bed introduction provides residence time for fines that would otherwise be entrained in the fluidizing gas and not converted in the bed.

The bed is usually preheated using an external burner fired by natural gas, propane or fuel oil. The hot flue gas from the external burner is used to heat the fluidized bed to the fuel ignition temperature. For biomass this temperature is around 540°C. Supplemental firing can be used to heat the freeboard gas to normal gasification temperature. At this point biomass is slowly introduced into the bed to raise the bed temperature to the desired operating range, normally 790-870°C. Bed temperature is governed by the desire to obtain complete devolatilization versus the need to maintain the bed temperature below the ash-fusion temperature of the biomass ash. As biomass is introduced into the bed, most of the organics pyrolytically vaporize and are partially combusted in the bed. The exothermic combustion provides the heat to maintain the bed at temperature and to volatilize additional biomass. Fluidized-bed gasifiers have the advantage of extremely good mixing and high heat transfer, resulting in very uniform bed conditions. Gasification is very efficient, and 95%-99% carbon conversion is typical. Bubbling fluidized-bed gasifiers are normally designed for complete ash carryover, necessitating the addition of cyclones for particulate control.

Circulating Fluid Bed

If the gas flow of a bubbling fluid bed is increased, the gas bubbles become larger, forming large voids in the bed and entraining substantial amounts of solids. This type of bed is referred to as a turbulent fluid bed (Babcock and Wilcox 1992). In a circulating fluid bed the turbulent bed solids are collected, separated from the gas, and returned to the bed, forming a solids circulation loop. A circulating fluid bed can be differentiated from a bubbling fluid bed in that there is no distinct separation between the dense solids zone and the dilute solids zone. Circulating fluid-bed densities are about 560 kg/m^3 compared to a bubbling-bed density of 720 kg/m^3 (Babcock and Wilcox 1992). To achieve the lower bed density, gas rates are increased from the 1.5-3.7 m/s of bubbling beds to about 9.1 m/s. The residence time of the solids in a circulating fluid bed is determined by the solids circulation rate, the attrition of the solids, and the collection efficiency of the solids separation device. Typical gas compoitions for fluid and circulating fluid bed gasifiers are shown in Table 2.

Table 2: Gas composition for fluid bed and circulating fluid bed gasifiers

Gasifier	FERCO	Carbona	Princeton Model	IGT
Type	Indirect CFB	Air FB	Indirect FB	PFB
Agent	steam	air	steam	O2/steam
Bed Material	olivine	sand	none	alumina
Feed	wood chips	wood pellets	black liquor	wood chips
Gas Composition				
H2	26.2	21.70	29.4	19.1
CO	38.2	23.8	39.2	11.1
CO2	15.1	9.4	13.1	28.9
N2	2	41.6	0.2	27.8
CH4	14.9	0.08	13.0	11.2
C2+	4	0.6	4.4	2.0
GCV, MJ/Nm^3	16.3	5.4	17.2	9.2

Entrained-flow

In entrained-flow gasifiers, pulverized feed is fed dry or in a slurry continuously into a pneumatic-flow reactor along with a relatively large amount of oxygen or steam. For oxidative heating the high temperatures caused by added oxygen almost completely destroy oils and tars. The high temperature (typically 1,300-1,400ºC) also means that the ash is typically removed as a liquid slag. These gasifiers have been developed for coal, and only very limited testing with biomass has been performed. Shell, Chevron Texaco, and Koppers-Totzek have developed entrained-flow reactors for coal. There are a number of reasons for the lack of application of these coal reactors to biomass, but the high cost of feed preparation to reduce moisture content to low levels and reduce the particle size, along with low heat content, is the primary concern (Larson and Katofsky 1992). Refractory life is also a concern for biomass feeds with high potassium content (Higman and van der Burgt 2003). Indirectly heated entrained-flow gasifiers have been developed for biomass by a number of companies, including Brightstar Environmental and Pearson Technologies.

System Applications

Biomass gasifiers can be used for many process applications, including process heat production, combined heat and power, stand-alone electricity production, and chemicals production. Technologies being commercialized today are primarily for thermal applications and electricity production. Development of commercial technologies for chemical synthesis is much further into the future.

References

Babcock and Wilcox (1992). Steam, 40th ed., Chapter 16: Atmospheric Pressure Fluidized-Bed Boilers, Babcock and Wilcox, Barberton, OH.

Graboski, M. and R.L. Bain (1979). "Properties of Biomass Relevant to Gasification," Chapter 3 of A Survey of Biomass Gasification - Volume 2, Solar Energy Research Institute, SERI/TR-33-239, National Renewable Energy Laboratory, Golden, CO, July.

Hansen, J.L. (1992). "Fluidized Bed Combustion of Biomass: an Overview," Biomass Combustion conference, Reno, Nevada, U.S. DOE Western Regional Biomass Energy Program, Jan. 28-30.

Higman, C. and M. van der Burgt (2003). Gasification, Elsevier, Amsterdam, ISBN 0-7506-7707-4, pg 147-154.

Larson, E.D. and R.E. Katofsky (1992). "Production of Hydrogen and Methanol from Biomass" Princeton University, Princeton, NJ, Report No. PU/CES 271, July.

Perry, R.H. and C.H. Chilton (1973). Chemical Engineers' Handbook, Fifth Ed., McGraw-Hill Book Co, New York.

Effective Thermal Design of Condensers for Mixtures

Thomas Lestina and R. Stanley Kistler
Heat Transfer Research, Inc., College Station, Texas

Introduction

Effective thermal design of condensers is required for energy efficient process design and operation. In general, methods to design heat exchangers for condensing mixtures have been established for decades, but new process applications require improvements in these methods. For example, in the process industry, the demand for increased production with low energy costs has resulted in reductions in the mean temperature difference (MTD) between process streams (such as for feed-effluent configurations), thus lowering the heat flux and available performance margin for condenser operation. In the power industry, cycles using binary mixtures (such as ammonia-water and hydrocarbons) are under evaluation as an alternative to the traditional steam cycle for lower temperature sources of energy. In the refrigeration industry, mixture refrigeration cycles are being evaluated for emerging applications such as solar powered air conditioning. In summary, cost effective designs require reliable thermal design practices for heat exchangers with condensing mixtures.

This tutorial discusses the fundamentals of mixture condensation and the key elements for a successful design.

Mixture Condensation Fundamentals

Figure 1 illustrates the fundamental principles of mixture condensation.

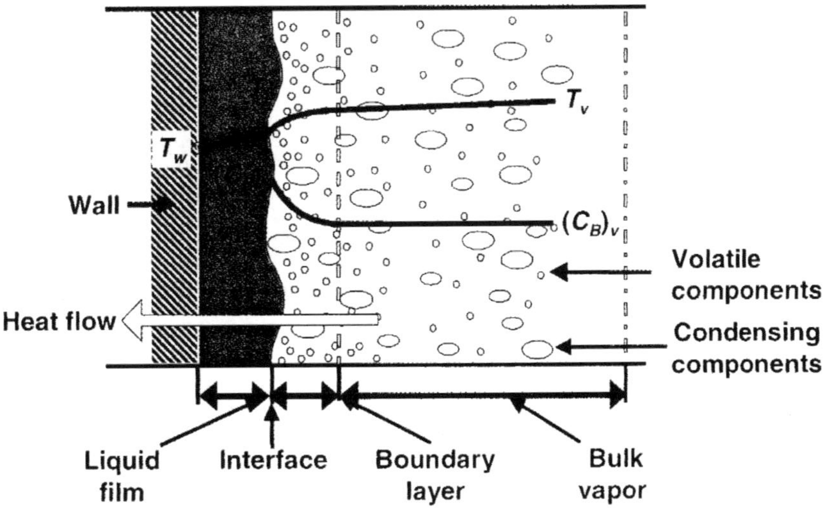

Figure 1. Schematic cross section of mixture condensation

The larger vapor components condense more readily than the smaller volatile components. Near the interface and within the boundary layer, the concentration of the volatile component, $(C_B)_v$, is higher than in the bulk vapor stream. This layer of high concentration of the volatile component results in a mass diffusion resistance which results in a temperature gradient from the bulk vapor to the interface. Rigorous calculation of this thermal gradient was originally proposed by Colburn-Hougen [1]. Because calculating the interface temperature is complicated, most practical heat transfer analyses eliminate the need to do so. A common method originally attributed to Bell-Ghaly [2], is summarized as follows:

$$T_V - T_W = T_V - T_i + T_i - T_W \tag{1}$$

$$\frac{q_t}{h_c} = \frac{q_{sv}}{h_{sv}} + \frac{q_t}{h_{cf}} \tag{2}$$

Rearranging Equation (2) allows the determination of the condensing coefficient on an incremental basis:

$$\frac{1}{h_c} = \frac{1}{h_{cf}} + \frac{q_{sv}}{q_t\, h_{sv}} \tag{3}$$

A basic shortcoming of the method is that it cannot handle mass diffusion effects properly. More sophisticated methods modify the duty ratio (q_{sv}/q_t) for diffusion effects. However, that requires a model for the mass diffusivity. The Lewis number (Le), which is the ratio of the mass diffusivity to thermal diffusivity, can be used to better understand the effect of mass diffusivity.

- For mixtures where the boundary layer for thermal gradient is approximately the same as the boundary layer for concentration gradient (as in Figure 1), the Lewis number is one.

- For mixtures where molecules of the condensing component are much larger than molecules of the volatile component, the Lewis number is less than one and the concentration boundary layer thickness is much greater than the thermal boundary layer thickness (the heat transfer coefficient increases as the Lewis number decreases).

- For mixtures where the condensing component is about the same size as the volatile component, the Lewis number is greater than one and the thermal boundary layer thickness is greater than the concentration boundary layer thickness.

In general, the concentration profiles for the condensing constituents are needed to calculate the variation in heat transfer coefficient.

Practical Guidelines for Design

Incremental Calculations

Condensers typically have large variations in the heat transfer coefficient and MTD along the condensing path. To accurately account for these variations, a calculation method which divides the condenser area into smaller incremental areas is needed. Designers should carefully review the capability of the software used to perform these calculations to ensure that it both accurately models the variations in MTD, condensate film coefficient, and vapor phase resistance and includes features to determine the condensing regimes and heat release data as discussed below.

Condensing Regimes

The condensate film coefficient changes significantly depending on the condensing regime. At the inlet, the vapor is often superheated. For desuperheating, there may be no condensate film or a very thin condensate film (for wet-wall desuperheating). As shown in Figure 2, once the bulk vapor reaches the dew point temperature, the condensate film is often quite thin, and the corresponding film coefficient is high because the condensate is stripped from the heat transfer surface. As the weight fraction vapor decreases, the velocity decreases and the condensate film drains by gravity, as also shown in Figure 2. In this gravity-controlled regime, the condensate film coefficient is often substantially less than in the shear-controlled regime because the thickness of the film is greater. An effective calculation method should account for this variation in condensate film coefficient due to condensing regimes.

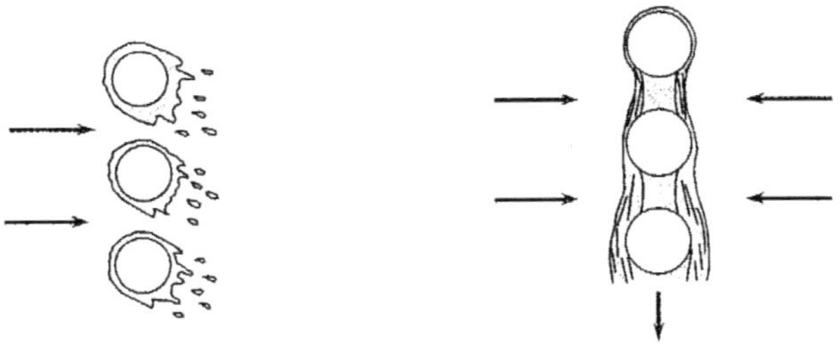

Figure 2. Shear- and gravity-controlled shellside condensation

Heat Release Curves

The heat release curve is the grid of duty/enthalpy, temperature, pressure, and weight fraction vapor for the condensing mixture. This heat release data is calculated based on the mixture composition and mixing rules. Accurate heat release data is essential to determining the incremental MTD. Furthermore, the gradient dT/dy is related to the heat flux ratio, q_t/q_{sv}, and the resistance in the vapor phase. In mixture condensation, the dT/dy gradient is usually low near the inlet where the heavier component readily condenses and high near the outlet where the lighter components are condensing.

Inert and Condensate Removal

Even though inert and/or condensate buildup are not included in the thermal design, adequate removal of any inert components and condensate ensures that the condenser will meet the required duty. Since inerts are swept along with the vapor and condensate, they should be removed near the outlet at the lowest pressure portion of the condenser. Also, buildup of condensate with substantial subcooling should be avoided for effective thermal designs.

References

1. Colburn, A. P., and Hougen, O. A., Design of Coolers/Condensers for Mixtures of Vapors with Non-condensing Gases, *I&EC*, Vol. 26, No. 11, p. 1178 – 1182, 1934.

2. Bell, K. J., and Ghaly, M. A., An Approximate Generalized Design Method for Multicomponent/Partial Condensers, *CEP Symp. Series*, Vol. 69, No. 72, 1973.

Nomenclature

$(C_B)_v$	Concentration of component *B* in vapor	—	—
$(C_B)_i$	Concentration of component *B* at vapor-liquid interface	—	—
h_c	Combined condensation heat transfer coefficient, including vapor-phase resistance	W/m^2 K	Btu/hr ft^2 °F
h_{cf}	Condensate film heat transfer coefficient	W/m^2 K	Btu/hr ft^2 °F
h_{sv}	Sensible heat transfer for vapor phase (including noncondensables) if flowing alone	W/m^2 K	Btu/hr ft^2 °F
Le	Lewis number	—	—
MTD	Mean temperature difference	°C	°F
q	Heat flux	W/m^2	Btu/hr ft^2
q_{sv}	Sensible vapor heat flux	W/m^2	Btu/hr ft^2
q_t	Total heat flux	W/m^2	Btu/hr ft^2
T	Temperature	°C	°F
T_i	Vapor-liquid interface temperature	°C	°F
T_v	Vapor temperature	°C	°F
T_w	Wall temperature	°C	°F
y	Weight fraction vapor	—	—

American Institute of Chemical Engineers
3 Park Avenue
New York, New York 10016-5991

ISBN: 978-1-60423-476-3